RACE AND THE GENETIC REVOLUTION

This book is a project of the Council for Responsible Genetics. Since 1983, the Council for Responsible Genetics has represented the public interest and fostered public debate about the social, ethical, and environmental implications of genetic technologies. CRG is a leader in the movement to steer biotechnology toward the advancement of public health, environmental protection, equal justice, and respect for human rights. (See www.councilforresponsiblegenetics.org.)

RACE AND THE GENETIC REVOLUTION

SCIENCE, MYTH, AND CULTURE

EDITED BY SHELDON KRIMSKY
AND KATHLEEN SLOAN

With a Foreword by Evelynn M. Hammonds

A Project of the Council for Responsible Genetics

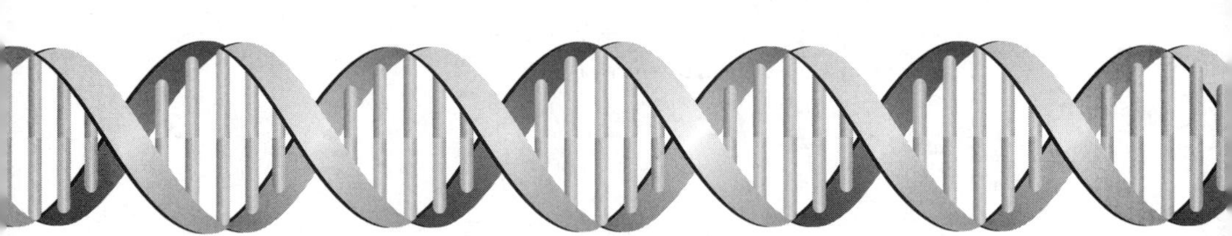

Columbia University Press *New York*

Columbia University Press
Publishers Since 1893
New York Chichester, West Sussex

cup.columbia.edu

The authors and Columbia University Press gratefully acknowledge the support of the Council for Responsible Genetics (www.councilforresponsiblegenetics.org) in the publication of this book.

Library of Congress Cataloging-in-Publication Data
Race and the genetic revolution : science, myth, and culture / edited by Sheldon Krimsky and Kathleen Sloan.
p. ; cm.
Includes index.
"A project of the Council for Responsible Genetics."
ISBN 978-0-231-15696-7 (cloth : alk. paper) — ISBN 978-0-231-15697-4 (pbk. : alk. paper) — ISBN 978-0-231-52769-9 (ebook)
1. Human population genetics. 2. Race. I. Krimsky, Sheldon.
II. Sloan, Kathleen, 1958- III. Council for Responsible Genetics
[DNLM: 1. Continental Population Groups—genetics. 2. Genetics.
3. Evolution, Molecular. 4. Genetic Variation. QU 450]

GN289.R33 2011
576.5'8--dc23 2011021226

Columbia University Press books are printed on permanent and durable acid-free paper.
This book is printed on paper with recycled content.
Printed in the United States of America

c 10 9 8 7 6 5 4 3 2 1
p 10 9 8 7 6 5 4 3 2 1

SK
Dedicated to those who have argued and fought against the use of the concept of "race" to create social hierarchies and perpetuate scientific myths.

KS
Dedicated to all those scholars and advocates for progressive, enlightened societal change who inspire me to persevere and blaze new trails.

CONTENTS

FOREWORD

by Evelynn Hammonds

For many readers the genetics/genomics revolution is one they know most about from newspaper accounts of the discoveries of new genes or the use of DNA for forensic analysis in high-profile criminal cases or from popular television shows. Among the general public and in the academy there has been little attention paid to the increasing scholarly work on the broader impact of the genetic revolution in American society.

The distinguished scholars in this volume note correctly that one of the most significant issues that these new genetic technologies engage with is that of race and racial disparities in the criminal justice system and in health care. Despite comments that suggest we now live in a post-racial moment here in the twenty-first century, the vexing, paradoxical, and ideological aspects of the race concept are still with us. How can this be? Didn't we learn that after the abuses of the race concept during World War II we entered an era where modern biological and biomedical sciences repudiated the idea that human beings could be meaningfully grouped into "races"? Didn't we recently learn with the completion of the sequencing of the human genome that "the concept of race has no genetic or scientific basis"? It is no surprise to the scholars writing in this volume that race remains a problem for genomics, biomedicine, and forensic sciences. This volume will serve as a critical corrective for scholars and the public to the view that race is no longer a useful concept for understanding human variation in the twenty-first century.

This book provides a careful debunking of common misperceptions about how race is used in the sciences under discussion. Most importantly, it shows how racial concepts change over time from earlier links to bodily differences or phenotypes to recent associations with variation in genotypes. These analyses go beyond positing race as a concept that is either social and historical *or* biological, but rather they explore how race is embedded in scientific practices that utilize mechanisms to sort people into groups for the purposes of classification and identification. Biological difference then becomes the substrate for existing social realities, and race becomes for some scientists a readily available, if imprecise and illogical, way to answer questions about the genetics of human variation.

This book does not question the fundamental validity of genetics for exploring differences between and within human populations. Rather the biologists, legal scholars, and social scientists represented in this volume provide cogent and insightful analyses that reveal how race continues to be a distorting mirror of both our society and new scientific developments in genetics.

Evelynn M. Hammonds, PhD

Barbara Gutmann Rosenkrantz
Professor of the History of
Science and of African and
African American Studies
and Dean of Harvard College

ACKNOWLEDGMENTS

This book was inspired by projects on genetics and race coordinated through the Council for Responsible Genetics (CRG) and funded by the Ford Foundation from 2007 through 2009. Without the support of Ford and CRG, this volume would not have been possible. In particular, we wish to thank Todd Cox and Kirsten Levingston, our Program Officers at the Ford Foundation.

We were fortunate in being able to assemble a distinguished group of contributors: Troy Duster, Duana Fullwiley, Joseph Graves, Jr., Elena L. Grigorenko, Jonathan Kahn, Kenneth K. Kidd, Osagie Obasogie, Pilar Ossorio, Robert Pollack, Michael Risher, Steven E. Stemler, Robert J. Sternberg, Helen Wallace, Patricia Williams, and Michael Yudell. These leading scholars found the time in their overwhelmingly busy schedules to meet our publishing deadline. We are indebted to them for their cooperation and enthusiastic support of this endeavor.

We are especially grateful for the contributions of Tania Simoncelli, who oversaw the planning and orchestration of a conference on race and forensic DNA databases that was held at New York University in June 2008 and resulted in the production of several draft contributions to this volume.

We are deeply appreciative of the assistance and support provided by Jeremy Gruber, president of the Council for Responsible Genetics, during the second phase of the Ford Foundation grant. We also wish to acknowledge the Board of CRG for its encouragement and support. Finally, we wholeheartedly thank our team at Columbia University Press, Patrick Fitzgerald and Bridget Flannery-McCoy, who shepherded the process of turning our manuscript into a published book, and Robert Demke for his copyediting.

RACE AND THE GENETIC REVOLUTION

INTRODUCTION

HOW SCIENCE EMBRACED THE RACIALIZATION OF HUMAN POPULATIONS

Sheldon Krimsky

This volume of essays grew out of two projects of the Council for Responsible Genetics (CRG) that examined the persistence of the concept of human races within science and the impacts such a concept has had on disparities among people of different geographical ancestries. Both projects were funded by the Ford Foundation. The first project commenced with a series of research papers addressing the effects of expanded forensic DNA databases on "racial" disparities in the criminal justice system and culminated in a national conference held at New York University on June 19–20, 2008, that brought together academics and social justice advocates to discuss "racialized" forensic DNA databases and explore policy solutions.

The second project explored the impacts of modern genetics on reinscribing and objectifying the concept of "race" in science and society. A series of papers were commissioned and subsequent forums were hold at the California Endowment's Oakland Conference Center at Oakland on August 13, 2009, and at the American Museum of Natural History in New York on August 20, 2009. The CRG assembled a multidisciplinary group of scholars and community activists to contribute papers that discussed the history of the concept of "race" and how the new field of genomics informs scientific, medical, and public understanding of "race" in new areas like forensic DNA, racialized medicine, and intelligence testing.

Other papers were contributed to this volume after the conferences were held. It has been the intention of the Council for Responsible Genetics and the editors of this volume to draw attention to myths about "race" and to bring public awareness to the impacts such myths and scientific misunderstandings can have in the pursuit of social equality for all people regardless of the color of their skin, their ethnic identify, or the geographical origin and phenotype of their ancestry.

Historically, the concept of "race" has been steeped in paradox, embraced by ideology, adopted and rejected by science, but nevertheless remains an indisputable part of public discourse. The term "race" is merely a shadow of what it once represented in science. Simply put, race is a scientific myth and a social reality.[1]

Rooted in the science of zoology, the concept of "race" was initially introduced as a taxonomic category for classifying the organizational structure of animal species. The organization of living things was classified into broad categories of kingdom, phylum, class, order, family, genus, species, where kingdom is the broadest category. The term "race" in zoology is applied in formal animal taxonomy to variations below the species level. Races are interbreeding groups of animals, all of whom are genetically distinct from the members of other such groups of the same species. However, these groups (races) are geographically isolated from one another, so there are barriers to genetic exchange between groups. In the eighteenth-century scientists began applying the term "race" to human populations.

In *The Descent of Man*, Charles Darwin noted the divergence among naturalists in deciding on the correct number of human races: "Man has been studied more carefully than any other animal, and yet there is the greatest possible diversity amongst capable judges whether he should be classed as a single species or race, or as two (Virey), as three (Jacquinot), as four (Kant), five (Blumenbach), six (Buffon), seven (Hunter), eight (Agassiz), eleven (Pickering), fifteen (Bory St. Vincent), sixteen (Desmoulins), twenty-two (Morton), sixty (Crawfurd), or as sixty-three, according to Burke."[2] Yet Darwin, like many scientists of his time, was not ready to discard the idea of race as having scientific legitimacy. He wrote, "this diversity of judgment does not prove that the races ought not to be ranked as species, but it shews that they graduate into each other, and that it is hardly possible to discover clear distinctive characters between them."[3]

The idea that there were fixed, unalterable human morphological or genetic qualities of certain population groups, transmitted from generation to generation, was in disfavor by most scientists in the late nineteenth century. Even the German physical anthropologist Johann Friedrich Blumenbach, who is often credited because of his book *On the Natural Variety of Mankind* as one of the progenitors of racializing the human population, acknowledged that no sharp distinction can be made between people.[4] He wrote, "No variety of mankind exists, whether of colour, countenance, or stature, etc., so singular as not to be connected with others, of the same kind by such an imperceptible transition that it is very clear that they are all related, or only differ from each other in degree."[5]

In this volume Michael Yudell provides an historical account of how "race" was embraced by science for over three centuries (see chapter 1). He shows us how a new generation of scientists in the early twentieth century began to apply modern genetics to an understanding of racial classification.

While the concept of "race" became redefined through population genetics, some scientists were preparing to exorcise it from science. In 1942 Ashley Montagu published *Man's Most Dangerous Myth: The Fallacy of Race.*[6] Expressing what he asserted was the consensus among scientists, Montagu wrote: "Most authorities of the present entertain no doubts as to the meaninglessness of the older anthropological conception of 'race.' They do not consider that any of the existing concepts of 'race' correspond to any reality whatever."[7]

Nearly sixty years later, Michael Omi reaffirmed the consensus: "Biologists, geneticists, and physical anthropologists among others, long ago reached a common understanding that race is not a 'scientific' concept rooted in discernable biological differences."[8] By the end of the twentieth century, race had been defined in the Unabridged Random House Dictionary as "an arbitrary classification of modern humans, sometimes, esp. formerly, based on any or a combination of various physical characteristics, as skin color, facial form, or eye shape."[9] Thus, "race" is a social construction. This is reflected in the words of Lisa Gannett, "The races biologists once claimed to have discovered in nature were, in actuality, the illegitimate offspring of an invented classification scheme they had imposed on nature."[10] In view of the consensus within science on "race," Keita and Kittles ask, "Why do the concepts of biological race and racial categories continue to exist and be utilized?"[11] This question leads us to a number of paradoxical elements, confusions, and public myths associated with how race continues to be used in science and the role it plays in social policy and popular culture.

It is not unusual for people to sort one another into group categories by external characteristics including ethnicity, language, skin color, and morphological features. These popular sorting mechanisms neglect genetic similarities as well as many nonobservable genetic differences. In the early 1970s scientists began studying how different population groups varied genetically. It became evident in the early 1970s, through a pathbreaking study by Richard Lewontin, that there was more genetic diversity within a population group (e.g., West Africans) than between two groups (West Africans and Europeans).[12] The genes for blood types do not divide up by geographical region, and genetic risks (disease polymorphisms) do not sort out according to the nineteenth-century meaning of race.[13] Keita and Kittles note, "'Race' is a legitimate taxonomic concept for chimpanzees but does not apply to humans (at this time)."[14]

Another source of paradox can be found in how "race" is currently used in the social sciences. In many areas of social science research, "race" is a variable. These include studies in income disparity, disease incidence, intelligence, and crime statistics to name a few. But since there are no biological or genetic markers for "race," it is operationally defined by self-identification. People sort themselves out as part of a survey according to which "race" they self-identify. But how a person self-identifies with a socially constructed idea of "race" depends on social and cultural norms. The same person may self-identify with one "race" in one country and another "race" in another country. Or a person may change his or her self-identification when the options before them have changed.

It must be understood and is rarely mentioned that research based on self-identification as the criteria for "race" correlates the independent variable (those who self-identify) with the dependent variable (e.g., bias in finding rental housing, poverty, illness). The social construction of "race" and its operational definition as "self-identity" means that resulting outcomes of the research will also be built on this social construction and not grounded in objective biological and social realities. And when the social definitions change, the significance of the research outcome will as well. Research results that use "race" defined by "self-identification" should be interpreted as "those who self-identify as race X compared to those who self-identify as race Y are more than twice as likely to be discriminated in housing."

For the 2010 United States Census, there were fourteen main categories of race offered in the questionnaire including a category of "some other race." Under "Other Asian" respondents are prompted to choose from among Laotian, Thai, Pakastani, Cambodian, or others. Someone of mixed race will undoubtedly feel inclined to choose from among the given categories.

This brings us into a second paradox. The term race as used in most contemporary societies is defined by certain discrete categories, currently fourteen "races" in the United States Census. Any physical trait or combination of traits that might be used in making such a classification or self-identification are continuous variables, representing centuries of genetic exchange (admixture) among populations. From a logical and mathematical standpoint, continuous variables cannot be mapped onto discrete variables. It makes as much nonsense to say that a light-skinned individual who has African ancestry is classified as "Black, African American, or Negro" as it would be to say that a dark-skinned person with any Northern European Ancestry is "White." On this point Stephen J. Gould wrote, "You cannot map a continuous distribution if all specimens must first be allocated to discrete subdivisions."[15]

So if it is scientifically unsound to conflate all of human genetic diversity into a few distinct racial types, why is it still done? One answer is that for some people in some contexts this sorting process serves some function, whether just or unjust. In the nineteenth century racial types were idealizations or constructs of a population group. Ingold writes: "Every race was thought to represent a type in the strict sense; that is to say, for each race there was supposed to correspond an essential form of the human being—Caucasoid, Mongoloid, Negro, and so on—to which every living individual represented a more or less close approximation."[16] People of mixed heritage, strictly speaking, didn't fit into the classification scheme. Today, multiracial individuals simply choose to associate themselves with a particular "racial group identity," from among choices that are imposed upon them, even if they are a poor approximation to the ideal type.

Of what value can racial categories, which are recognized as unscientific, arbitrary, socially constructed ideal types, have? If the dominant culture sorts and discriminates against people associated with one of these "ideal types," then this is of importance to social scientists and policy makers. Science can play a role in seeking to understand why this sorting and the consequences that derive from it takes place. But drawing generalizable conclusions in psychology, education, or medicine from these ideal types is baseless. The concept of race is a vestigial cultural artifact that persists in people's minds and public policies. The American Association of Physical Anthropology issued a consensus statement on race in 1996 that dismisses the idea of an average "racial type":

> Generally, the traits used to characterize a population are either independently inherited or show only varying degrees of association with one another within each population. Therefore the combination of these traits in an individual very commonly deviates from the average combination in the population. This fact renders untenable the idea of discrete races made up chiefly of typical representatives. . . . On every continent there are diverse populations that differ in language, economy and culture. There is no national, religious, linguistic or cultural group or economic class that constitutes a race.[17]

Given the preponderance of the science disavowing the classification of humans into races, how has this influenced scientific textbooks. Ann Morning studied how the concept of race was treated in biology textbooks over a period of fifty years, from 1952 to 2002. The results of her investigation were reported in the *American Journal of Sociology*. Morning found that "race appears to be returning, not disappearing, as a topic of biological instruction" and that "the textbooks' conceptual framing of race has changed markedly over the period from a model based on phenotype to one grounded on genotype."[18]

The reinscription of race into science is in large measure a consequence of the use of genetics in ascertaining the ancestry of individuals. This involves the development of genetic markers called Ancestry Informative Markers (AIMs): "An individual's African ancestry is inferred from genetic similarity along a few dozen genetic markers . . . derived from a few dozen cell lines from Central-West Africa, carefully chosen to be maximally different from a comparable sample of East Asians and Northern Europeans."[19] Jonathan Marks observes, "While this is not race in any previously familiar sense of the term it is readily conflated with such notions, whether ingeniously or not."[20]

The essays in this volume explore both historical and contemporary views of race and genetics. After the introduction, which discusses the vestigial remains of race in science, its undeniable socially constructed reality, and the paradoxes it leaves us, the book is divided into six sections with two essays per section. In part 1, "Science and Race," Michael Yudell provides a short history of the race concept covering the eighteenth to the twentieth century. Yudell shows us how science, ideology, and policy become intertwined over the concept of race. Robert Pollack provides a geneticist's view of the idea of race in the context of natural selection and human evolution.

In the part titled "Forensic DNA Databases, Race, and the Criminal Justice System" Michael Risher discusses US DNA databanks that contain the forensic DNA profiles of individuals of interest to local, state, and federal criminal justice systems. He examines the problem of racial disparities resulting from policies on obtaining DNA profiles. Helen Wallace contributes a parallel analysis of racial disparities in forensic DNA databanks in the United Kingdom, where children ten years old and over may have their DNA placed on a national database.

In part 3, "Ancestry Testing," Troy Duster and Duana Fullwiley each contribute essays that examine the assumptions behind the current methods of tracing the percentage ancestry of individuals to a few major regions of the world. The industry that has developed around "finding your ancestry" has reinscribed race into a scientific vernacular.

Jonathan Kahn and Joseph L. Graves each contribute essays in part 4, "Racialized Medicine," where they discuss the assumptions behind the decision by the US Food and Drug Administration to approve a cardiac drug (BiDil) specifically for people of African American ancestry. Kahn discusses how BiDil was developed and approved for use by a single "racial" group. Graves examines the fallacies of racialized medicine and how an evolutionary approach to medicine can contribute toward the elimination of health disparities.

In part 5, "Intelligence and Race," Pilar Ossorio critically examines hereditarian theories of intelligence and discusses the implications of the view that IQ testing by "race" tells us anything about how innate intelligence is distributed among population groups. Sternberg, Grigorenko, Kidd, and Stemler provide a comprehensive review and analysis of the scientific literature on intelligence, race, and genetics. They explore and evaluate claims made about the alleged relationships between race and intelligence through the lens of genetics.

Finally, part 6, "Contemporary Culture, Race, and Genetics" contains contributions by Patricia Williams on cultural views about race and by Osagie K. Obasogie, who introduces the idea of racial impact assessments. Williams explores how the concept of race has been incorporated into personal narratives of identity. Obasogie examines the consequences of an intractable racialization of society and its impact on minority populations. His concept of racial impact assessments seeks to address the disparities that arise from socially constructed ideas of race. In the conclusion, Kathleen Sloan discusses the educational responsibility of science to advance social justice in a world where race retains a powerful social meaning, albeit one that varies significantly across cultures.

NOTES

1. In answering the question of why "race" retains such force in society, Adolph Reed writes, "The short answer is that race is a social reality." Adolph Reed, "Making Sense of Race, I: The Ideology of Race, the Biology or Human Variation, and the Problem of Medical and Public Health Research, "*Journal of Race and Policy* 1 (2005):11–42, at 13.
2. Charles Darwin, *The Descent of Man* (New York: The Modern Library), 536. John Joseph Virey (1774–1847); Honoré Jacquinot (1815–1887); Immanuel Kant (1724–1804); Johann Friedrich Blumenbach (1752–1840); Georges-Louis Leclerc, Compte de Buffon (1707–1788); John Hunter (1728–1793); Louis Agassiz (1807–1873); Charles Pickering (1805–1878); Jean Baptiste Bory de Saint Vincent (1778–1846); Charles Robert Alexandre des Moulins (1798–1875); Samuel George Morton (1799–1859); John Crawfurd (1783–1868); Edmund Burke (1729–1797).
3. Darwin, *The Descent of Man*, 536.
4. Johann F. Blumenbach, *On the Variety of Mankind*, trans. T. Bendyshe (London: Anthropological Society, 1865).
5. Ashley Montagu, *Man's Most Dangerous Myth: The Fallacy of Race* (New York: World Publishing Co., 1964), 41. Originally published in 1942 by Columbia University Press.
6. Ibid.
7. Ibid., 130.

8. Michale Omi, "The Changing Meaning of Race," in *America Becoming: Racial Trends and Their Consequences*, ed. Neil J. Smelser, William J. Wilson, and Faith Mitchell (Washington, DC: National Academy Press, 2001), 243
9. Random House Unabridged Dictionary, 1993.
10. Lisa Gannett, "The Biological Reification of Race," *British Journal of the Philosophy of Science* 55 (2004): 323–75, at 323.
11. S. O. Y. Keita and Rick A. Kittles, "The Persistence of Racial Thinking and the Myth of Racial Divergence," *American Anthropologist* 99 (1997): 534–44, at 511.
12. Richard C. Lewontin, "The Apportionment of Human Diversity," *Evolutionary Biology* 6 (1972): 381–98.
13. Jonathan Marks, "Race: Past, Present, and Future," in *Revisiting Race in a Genomic Age*, ed. Barbara A. Koenig, Sandra Soo-Jin Lee, and Sara S. Richardson (New Brunswick, NJ: Rutgers University Press, 2008), 21–38, at 34
14. S. O. Y. Keita, R. A. Kittles, C. D. M. Royal et al, "Conceptualizing Human Variation," *Nature Genetics* 36 (2004): S17–S20, at S19.
15. Stephen J. Gould, "Why We Should Never Name Race," in *Ever Since Darwin* (New York: Norton, 1977), 235.
16. Tim Ingold, "When Biology Goes Underground: Genes and the Spectre of Race," *Genomics, Society and Policy* 4 (2008): 23–37, at 25.
17. American Association of Physical Anthropology, "AAPA Statement on Biological Aspects of Race," *American Journal of Anthropology* 101 (1996): 569–70, at 570.
18. Ann Morning, "Reconstructing Race in Science and Society: Biology Textbooks, 1952–2002," *American Journal of Sociology* 114 (2008): S100–S137, at S119.
19. Marks, "Race," 25.
20. Ibid.

PART I

SCIENCE AND RACE

Historical and Evolutionary Perspectives

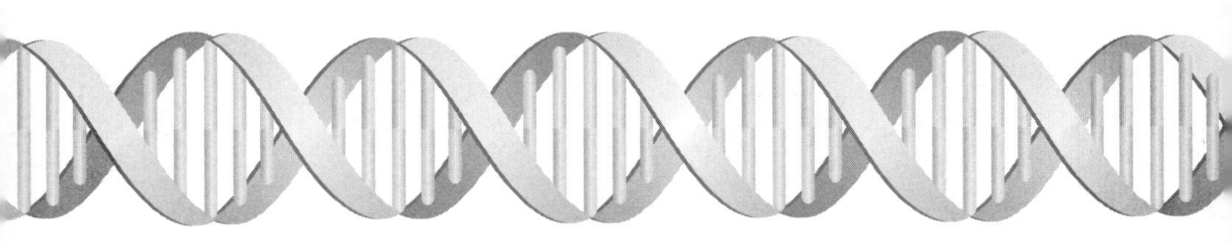

1

A SHORT HISTORY OF THE RACE CONCEPT

Michael Yudell

At the dawn of the twenty-first century, the idea of race—the belief that the peoples of the world can be organized into biologically distinctive groups, each with their own physical, social, and intellectual characteristics—is understood by most natural and social scientists to be an unsound concept. The way scientists think about race today, after all, is different than it was in the wake of the Civil Rights Movement of the 1950s and '60s when some promoted black genetic inferiority as an argument against egalitarian social and economic policy, and certainly different than one or two centuries ago as scientific justifications for slavery and later Jim Crow were articulated. In other words, race, its scientific meaning seemingly drawn from the visual and genetic cues of human diversity, is an idea with a measurable past, identifiable present, and uncertain future. These changes are influenced by a range of variables including geography, politics, culture, science, and economics.

Today, despite the growing consensus among scientists that race is not, in fact, a useful classificatory tool, an understanding of human difference and diversity remains a hallmark of contemporary scientific practice, and thus presents a seeming contradiction—how can one study human difference without talking about race?[1] On the one hand, beginning in the 1930s, advances in population genetics and evolutionary biology led many to conclude that the race concept was not a particularly useful or accurate marker of biological difference. By the 1970s, many prominent biologists came to see the race concept as a deeply flawed way to organize human genetic diversity that is inseparable from the social prejudices about human difference that spawned the concept in the eighteenth century and have accompanied its meaning since.[2] Historians and social scientists believe that race is socially constructed, meaning that the biological meaning of race has been constrained by the social context in which racial research has taken place.

On the other hand, because studying genetic differences can improve our understanding of human evolution, disease, and development, the relationship between genetics and human diversity remains an ongoing area of scientific inquiry. The challenge has been to develop a new scientific terminology and methodology that finds meaning in the study of human difference without recapitulating outmoded and racist notions often associated with the concept of race itself. Some scientists have developed novel ways to measure difference between various human populations, including using ancestry, ethnicity, and population as replacements or surrogates for race. Others, however, remain steadfast in their belief that technological and methodological improvements now allow an examination of racial difference with increasing precision that is disconnected from any social prejudices.

This chapter describes the role that scientific thought has played, from the late eighteenth through the twentieth century, in developing a language to measure the meaning of human difference in the form of race, and will also describe how over the course of the twentieth century, many scientists came to reject this concept. Finally, this chapter concludes with a brief examination of the current state of racial thinking in biology.

PRE-TWENTIETH-CENTURY CONCEPTIONS OF RACE

Ever since Thomas Jefferson wrote in the Declaration of Independence "that all men are created equal," America has struggled with the chasm between this Jeffersonian ideal and the realities of the American experience. Jefferson himself was the author of some of America's earliest ideas about race and science. In 1787, little more than a decade after he had penned the Declaration, Jefferson suggested in his work on the natural history of Virginia, *Notes on the State of Virginia*, that the difference between the races "is fixed in nature," and hypothesized that blacks were "originally a distinct race."[3] The contradiction between the Declaration and *Notes* may be understood, however, by Jefferson's view of humanity itself. If blacks were of a separate creation and set apart from the definition of "all men," then the equality set out in the Declaration did not apply to all.

Notwithstanding Jefferson's prominent voice on this issue (notably filled with contradictions, including longstanding evidence of a relationship and of at least one child with his slave Sally Hemings),[4] Americans, and before them, their European counterparts had long considered the nature of human difference. Historian Frank Snowden, looking at black-white contact before the sixth century C.E. found that although there is an "association of blackness with ill omens, demons, the devil, and sin, there is in the extant record no stereotyped image of Ethiopians as the personification of demons or the devil."[5] In ancient Greece and Rome "the major divisions between people were more clearly understood as being between the civic and the barbarous," between the political citizen and those outside of the *polis*, and not between bloodlines or skin color.[6] Most scholars now accept the viewpoint that in the ancient world "no concept truly equivalent to that of 'race' can be detected in the thought of the Greeks, Romans, and early Christians."[7] Rooting human variation in blood or in kinship was a relatively new way to categorize humans. The idea gained strength toward the end of the Middle Ages as anti-Jewish feelings, which were rooted in an antagonism toward Jewish religious beliefs, began to evolve into anti-Semitism. These blood kinship beliefs rationalized anti-Jewish hatred instead as the hatred of a people. For example, Marranos, Spanish Jews who had been baptized, were considered, by virtue of their ancestry, a threat to Christendom because they could not prove purity of blood to the Inquisition.[8]

Beginning in the eighteenth century, at the height of the Age of Enlightenment in Europe, these ideas were applied to explaining the diversity of humankind, driven in part by the experiences with new peoples during colonial exploration, the need to rationalize the inferiority of certain peoples as slavery took hold in European colonies, and the development of a new science to assess and explain diversity in *all* species. While the term race existed before the eighteenth century, mostly to describe domesticated animals, it was introduced into the sciences by the French naturalist Louis LeClerc, Comte de Buffon, in 1749. Buffon saw clearly demarcated distinctions between the human races that were caused by varying climates. Buffon's climatological theory of difference was infused with notions of European superiority. To Buffon, the natural state of humanity was derived from the European, a people he believed "produced the most handsome and beautiful men" and represented the "genuine color of mankind."[9]

The Swedish botanist and naturalist Carolus Linnaeus also made lasting contributions to the race concept at this time. Linnaeus's "natural system," which became the basis for the classification of all species, divided humanity into four groups: *Americanus, Asiaticus, Africanus, and Europeaeus.* To these groups he ascribed typological, or observable physical and behavioral, characteristics. *Americanus* were "reddish, choleric, and erect; hair black . . . wide nostrils . . . obstinate, merry, free . . . regulated by customs." *Asiaticus* were "melancholy, stiff; hair black, dark eyes . . . severe, haughty, avaricious . . . ruled by opinions." *Africanus* were "black, phlegmatic . . . hair black, frizzled . . . nose flat; lips tumid; women without shame, they lactate profusely; crafty, indolent, negligent . . . governed by caprice." Finally, *Europeaeus* were "white, sanguine, muscular . . . eyes blue, gentle . . . inventive . . . governed by laws."[10] Toward the end of the eighteenth century, German scientist Johann Blumenbach constructed a racial classification that built upon Linnaeus's work and proposed five racial types: Caucasian, Mongolian, Ethiopian, American, and Malay. Blumenbach's addition posited the Caucasian as the ideal, or mean, race, and on either side of that mean were racial extremes, the Mongolian and Ethiopian on one side and the American and Malay on the other. Both divergences from the Caucasian ideal were considered inferior.[11]

If racial science is science employed for the purpose of degrading a people both intellectually and physically, then beginning in the nineteenth century, American scientists played an increasingly active role in its development, all the while shaping the race concept. Scientists such as Samuel Morton, Josiah Nott, and George Gliddon offered a variety of explanations for the nature of white racial superiority meant to address the nature of physical and intellectual differences between races, the "natural" positions of racial groups in American society, and the capacity for citizenship of nonwhites. At the core of this work, known as the American School of Anthropology, was the theory of polygeny—the belief that a hierarchy of human races had separate creations. Samuel Morton's experiments on cranial capacity and intelligence sought to demonstrate this theory. Morton collected hundreds of skulls from around the globe, measured their volume, and concluded that the Caucasian and Mongolian races had the highest cranial capacity and thus the highest levels of intelligence, while Africans had the lowest cranial capacity and thus the lowest levels of intelligence. This work became the basis for more than a century of work studying intelligence and race. More than a century after Morton's death, the evolutionary biologist Stephen Jay Gould, using Morton's raw data, could not replicate the earlier findings. Gould concluded that Morton's subjective ideas about race difference influenced his methods and conclusions, leading to the omission of contradictory data and to the conscious or unconscious stuffing or underfilling of certain skulls to match his preordained conclusions. Indeed, the case of Samuel Morton illustrates how social conceptions of human difference shape the science of race. A recent paper argues Gould's criticisms of Morton's methodology "are poorly supported," calling into question the accusation that Morton fabricated data. Regardless, Morton's role in the development of the American School of Anthropology and the racist theory of polygeny, and the impact of these ideas on the study of race, are not in dispute and are strongly supported by the historical record.[12]

EARLY-TWENTIETH-CENTURY IDEAS ABOUT RACE

At the dawn of the twentieth century, explanations of racial difference based on measurable and observable physical traits such as cranial capacity and skin color gave way to a whole new way of thinking about the subject. Race instead came to be understood as a reflection of unseen differences that the scientists of the time attributed to the recently discovered factors of heredity, also known as genes. Genetics quickly came to provide the formative language of modern racism as ideas about racial differences became rooted in biology. This geneticization of race—the idea that racial differences in appearance and complex social behaviors can be understood as genetic distinctions between so-called racial groups—was shaped, in large part, by the eugenics movement. For the first two decades of the twentieth century, the disciplinary boundary between eugenics and genetics was fluid. Eugenics, according to its founder, the English scientist Francis Galton, promised to give "the more suitable races or strains of blood a better chance of prevailing over the less suitable."[13] This could be done either through positive eugenics, whereby certain groups were encouraged to breed with one another, or negative eugenics, whereby certain groups or individuals would be denied the right to reproduce, either through sterilization, as was the case in the United States, or through genocide, as was the case in Nazi Germany. Under the guise of this biological banner, eugenic racial science exerted a diverse influence, becoming a powerful ideological force in Nazi Germany, influencing the creation of eugenic sterilization laws in the United States that resulted in at least thirty thousand sterilizations, stoking racial hatred in early-twentieth-century America, and becoming a scientific buttress of twentieth-century American racial ideology.[14]

During the first three decades of the twentieth century, eugenicists and many geneticists fiercely advocated "the belief that human races differed hereditarily by important mental as well as physical traits, and that crosses between widely different races were biologically harmful." American eugenicists dedicated considerable resources to the study of black-white differences during the first three decades of the twentieth century, and sought to apply these ideas to the public sphere. Well-respected geneticists wrote openly that "miscegenation can only lead to unhappiness under present social conditions and must, we believe, under any social conditions be biologically wrong."[15] In his seminal work on race and intelligence, *Race Crossing in Jamaica*, Charles Davenport, a Harvard trained biologist and the titular head of the American eugenics movements from the outset of the twentieth century until the 1930s, wrote, "we are driven to the conclusion that there is a constitutional, hereditary, genetic basis for the difference between the two races [whites and blacks] in mental tests. We have to conclude that there are racial differences in mental capacity."[16] In their influential text *Applied Eugenics*, eugenicists Paul Popenoe and Roswell Hill Johnson, who endorsed segregation as a "social adaptation," wrote "that the Negro race differs greatly from the white race, mentally as well as physically, and that in many respects it may be said to be inferior when tested by the requirements of modern civilization and progress." Moreover, they suggested that "negroes, both children and adults, have been found markedly inferior to white in vital capacity. . . . Differences in temperament and emotional reaction also exist, and may be more important than the purely intellectual differences."[17] It must be stated that the genetic claims of racial difference advocated by eugenicists—from differences in intelligence to disease rates to musicality—have all been shown to be false.

Eugenic propagandists gave race an unalterable permanence; neither education, nor change in environment or climate, nor the eradication of racism itself could alter the fate of nonwhites. In the United States, the impact of eugenics on matters of human difference was felt widely. In Virginia, as head of the state's Bureau of Vital Statistics, eugenicist and white supremacist Walter Plecker helped to shape the state's segregation policies. For example, Plecker helped push Virginia's antimiscegenation Racial Integrity Act of 1924, and used that law to expose individuals he believed were passing as white in an attempt to stop what he feared to be the mongrelization of the races.[18]

African American intellectuals were prominent among those who responded to the growing chorus of scientific racist thinking at this time. In 1909 Kelly Miller, the Dean of Howard University, argued against scientific racism, writing that "since civilization is not an attribute of the color of skin, or curl of hair, or curve of lip, there is no necessity for changing such physical peculiarities."[19] The most determined critic of the biological race concept was W. E. B. Du Bois, a founder of the NAACP and editor of its magazine, *The Crisis*. Du Bois challenged the biological race concept at a time when science was being exploited in the service of racist ideas and practices. Du Bois was the first to synthesize a growing anthropologic literature that argued that race was not, in fact, a useful scientific category, and showed, instead, that race was socially constructed. For example, Du Bois believed race an ineffective measurement, given that "the human species so shade and mingle with each other that . . . it is impossible to draw a color line between black and other races."[20]

MID-TWENTIETH-CENTURY IDEAS ABOUT RACE

Beginning in the 1930s, an increasing number of geneticists, anthropologists, and social scientists began moving away from typological and eugenic descriptions of human difference to view races through the lens of population genetics and evolutionary biology. This approach rejected a eugenic notion of fixed genetic differences between so-called racial groups, and instead understood human races as dynamic populations distinguished by variations of the frequency of genes between populations. By rooting the meaning of race in genetic variation it became more difficult (though still possible) to argue that one race or another had particular traits specifically associated with it, or that one individual was typical of a race. Furthermore, the four or five racial groups identified by eighteenth- and nineteenth-century scientists, now varied depending upon the genes and traits examined by geneticists. Theodosious Dobzhansky, the evolutionary biologist whose work between the 1930s and 1970s had a tremendous influence on the way that scientists thought about race, concluded that the number of human races was variable depending upon what traits were being examined. In fact, Dobzhansky believed the concept of race in the context of population genetics and evolutionary biology was simply a tool for making genetic "diversity intelligible and manageable" in scientific study.[21] In other words, while human differences are real, the way we choose to organize those differences is a methodological decision and not one that reflects an underlying evolutionary hierarchy or the conservation of racialized traits through the admixture of populations. This new approach was brought about by new findings in genetics which demonstrated that genetic variation was much more common within species than once thought, and by the development of what is known as the evolutionary synthesis in biology, a union of population genetics, experimental genetics, and natural history that rejected eugenic notions of difference between and among species. Finally, changes in the concept of race were influenced by a growing cadre of scientists who were generally more liberal on matters of race than had been their predecessors.

From the 1930s through the 1950s, books by the biologists Theodosius Dobzhansky and L. C. Dunn, by the anthropologists Ruth Benedict and Ashley Montagu, by the political scientist and later Nobel laureate Ralph Bunche, and by the historian Jacques Barzun popularized the idea that race was not the immutable constant once proclaimed so by science.[22] Advances in genetics, particularly the discovery of the structure of the double helix in 1953, confirmed the complexity of human heredity and continued to undercut the simplistic theories of eugenicists and other racial scientists who advanced the idea of a fixed racial taxonomy. Yet, despite the best intentions by scientists like Dobzhansky and Dunn to reconceptualize the concept of race for modern biology, evidence suggests that these geneticists and their scientific allies ultimately helped to preserve the concept of race in science, and hence for use by both scientific and nonscientific racists—its methodological utility to evolutionary biologists and population geneticists would quickly be exploited and manipulated by these racists. Dobzhansky understood and feared this possibility. He acknowledged the imprecise nature of the race concept and worried that a genetic race concept could also begin to "serve as a racial standard with which individuals and groups of individuals can be compared" in the same way that a typological concept of race could.[23] But for him and other population geneticists and evolutionary biologists at midcentury, the concept of race was a methodological tool by which to measure genetic difference within species, not a way to understand the physical and intellectual differences between peoples with varying skin color.

The impact of this new way of thinking about race quickly made its way beyond scientific circles. For example, in 1944 *An American Dilemma: The Negro Problem and Modern Democracy*, by the Swedish economist Gunnar Myrdal, sought to recast America's racial problems as a moral conflict between the egalitarian impulses of America's democratic creed and its racist practices. Myrdal rejected the idea of white over black in an unchangeable biological hierarchy of races; his conclusion was influenced by changes in the biological race concept. Several chapters of *An American Dilemma* examined then-contemporary discoveries in genetics that led to the rejection of typological and eugenic notions of race in favor of race as "quantitative notions of the relative frequency of common ancestry and differentiating traits." Myrdal acknowledged "the great variability of traits among individuals in every population group . . . and the considerable amount of overlapping between all existing groups." Finally, Myrdal believed through genetics "the fundamental unity and similarity of mankind . . . is becoming scientifically established."[24]

It is significant to note that Myrdal's text is cited in the 1954 landmark US Supreme Court decision *Brown v. the Board of Education*, which unanimously struck down legalized segregation in public education. *Brown* did not comment directly on the nature of race or on the alleged superiority or inferiority of racial groups, yet, by identifying segregation's harmful impact on black children's psyches and the wrongness of causing these children "a feeling of inferiority," the Court implicitly acknowledged that thinking about races as inferior and superior was erroneous. Although not cited in the *Brown* ruling, it is hard to imagine that the publication of two United Nations Educational, Scientific, and Cultural Organization (UNESCO) "Statements on Race" in 1950 and 1951, both of which sought to place the race concept squarely in the context of population genetics and evolutionary biology, did not also have some impact on the thinking of the Court. The First Statement argued that "from the biological standpoint, the species *Homo sapiens* is made up of a number of populations, each one of which differs from the others in the frequency of one or more genes," and that "for all practical social purposes 'race' is not so much a biological phenomenon as a social myth." Although the First Statement, chaired by Ashley Montagu, called for abandoning the concept of race in favor of ethnicity, the Second Statement held fast to the validity of the population-genetics-based race concept as discussed above. Montagu, however, was not deterred, and spent much of the rest of his career fighting against the use of race in scientific thought, believing that its use was not scientifically appropriate.[25]

Even as biological and anthropological thought embraced the new population-genetics-based race concept, many scientists held fast to obsolete notions of race, suggesting that even widely accepted and validated science could not be an antidote to the racism of many in the field and beyond. R. A. Fisher, one of the founders of population genetics, asserted in 1951, for example, that "available scientific knowledge provides a firm basis for believing that groups of mankind differ in their innate capacity for intellectual and emotional development."[26] Writing in 1961, Carlton Coon, recently elected as President of the American Association of Physical Anthropologists, resuscitated the nineteenth-century scientific racism of Samuel Morton, arguing in his book *The Origin of Races* that the five races of humanity had separately evolved into *Homo sapiens*. Coon's confidant and cousin was the notorious mid-twentieth-century racist Carlton Putnum, whose racist tract *Race and Reason* drew heavily on the ideas in *The Origin of Races*.[27]

That these arguments about the nature of race and racial difference were occurring during the years of America's Civil Rights Movement help illustrate the relationship between science and society. Dobzhansky and other biologists remade the race concept because they believed it methodologically important to their work and because they wanted to jettison from scientific thought and practice the racism inherent in typological and eugenic ideas about race. Even as science debated the meaning of race in biology, the abhorrent effects of American racism were in plain view. The same week that UNESCO published its "Second Statement on Race" a mob of three thousand whites prevented an African American US Army veteran from moving into an apartment in a formerly all-white apartment building in Cicero, Illinois.[28]

LATE-TWENTIETH-CENTURY IDEAS ABOUT RACE

By the 1960s and 1970s, geneticists were able to reveal with increasing sophistication and precision the shortcomings of the concept of race in biology. In 1972, the geneticist Richard Lewontin, who had been a student of Dobzhansky's at Columbia in the 1950s and was considered a leader in his field, published a study showing that human populations were even more genetically diverse than once thought. Lewontin, using molecular genetic techniques in gel electrophoresis he himself had pioneered in the mid-1960s, found that most genetic variation (85.4 percent) was "contained within" racial groups or "between populations within a race" (8.3 percent), whereas only 6.3 percent of "human variation was accounted for by racial classification." From these findings, Lewontin concluded that race had "virtually no genetic . . . significance." After all, if more genetic diversity occurred within so-called racial groups than between them, then what exactly would race be measuring if it were meant to organize populations by genetic difference? Lewontin concluded that the "use of racial categories must take its justifications from some other source than biology. The remarkable feature of human evolution and history has been the very small degree of divergence between geographical populations as compared with the genetic variation among individuals."[29]

At the end of the twentieth century, the geneticist L. Luca Cavalli-Sforza confirmed Lewontin's findings using contemporary DNA techniques. His results showed that there was no significant genetic discontinuity between any so-called races in our species that would justify the use of racial classification in humans. Cavalli-Sforza believed that these results and the results of other studies implied that population genetics and evolutionary biology had satisfactorily shown that the "subdivision of the human population into a small number of clearly distinct, racial or continental, groups . . . is not supported by the present analysis of DNA." Given that studies had now confirmed Lewontin's results for almost three decades, Cavalli-Sforza believed that "the burden of proof is now on the supporters of a biological basis for human racial classification."[30]

Yet even as it became increasingly clear that the race concept was not a useful classificatory tool, several high profile scientists, none of whom were geneticists, continued to make claims that race was, in fact, a legitimate biological concept, and that those who argued against race had political, not scientific, agendas. Generally, it was from these findings that the public became aware of the ongoing debates about the nature of race in science. For example, in 1969 the educational psychologist Arthur Jensen, a professor at the University of California, Berkeley, argued that intelligence, or IQ, had high genetic heritability, and that therefore redress for racial discrepancies in IQ through education was useless.[31] A few years earlier the Nobel Prize winning physicist William Shockley, a professor at Stanford University and coinventor of the transistor, made similar claims, calling for the National Academies of Science to investigate the genetic aspects of what he called our nation's "slum problem."[32] The attempt by men like Jensen and Shockley to employ the biological race concept demonstrates that no matter how hard biologists like Dobzhansky, Montagu, and Lewontin tried to either narrowly define race in the context of biology or abandon it altogether, and that despite the stated shift away from typology to populationist thinking, race could and would be used for typological, racist, and nonscientific ends.

Rather than debate the biological nature of race, in the 1970s some scientists instead began to debate the biological nature of racism. Sociobiology, as developed by the entomologist E. O. Wilson, claimed racism, xenophobia, and ethnocentrism to be biological traits. Sociobiology offered theories on why, in an evolutionary and genetic way, populations of peoples hated, feared, and distrusted one another. So while the biological race concept was largely ignored by sociobiological theory, the social meanings of race became a focus of this research through the study of racism. In the wake of the social transformation brought about by the

Civil Rights Movement, the 1970s witnessed a backlash against the struggle for equality in the United States. A sociobiological theory of racism and its popularization seemed well suited to these times. As Henry Louis Gates has argued, sociobiology, in a sense, "naturalized racism" by providing a genetic basis for it.[33]

CONCLUSION: THE RACE CONCEPT IN THE TWENTY-FIRST CENTURY

At a June 2000 Rose Garden ceremony, President Bill Clinton, flanked by genome sequencers Francis Collins and Craig Venter, announced the completion of a draft sequence of the human genome, the complete sequence of human DNA. Collins, then head of the National Human Genome Research Institute, and Venter, then President of Celera Genomics, offered their genomic data to the world that is enhancing our understanding of human biology, and, in turn, will help public health and medical professionals prevent, treat, and cure disease.

On that day Venter and Collins emphasized that their work confirmed that human genetic diversity cannot be captured by the concept of race, and also showed that all humans have genome sequences that are 99.9 percent identical. At the White House celebration Venter said, "the concept of race has no genetic or scientific basis."[34] A year later, Collins wrote, "those who wish to draw precise racial boundaries around certain groups will not be able to use science as a legitimate justification."[35] Yet, since the White House announcement there has been an increase in claims that race is a biologically meaningful classification. Genetic epidemiologist Neil Risch believes that "identifying genetic differences between races and ethnic groups . . . is scientifically appropriate." Risch claims that race is essential to help determine "differences in treatment response or disease prevalence between racial/ethnic groups" and strongly supports the "search for candidate genes that contribute both to disease susceptibility and treatment response, both within and across racial/ethnic groups."[36] Even Collins himself suggested in 2005 that we now need to study how genetic variation and disease risk are correlated with "self-identified race, and how we can use that correlation to reduce the risk of people getting sick."[37] And this is the very paradox of the genomic era when it comes to race; with renewed precision scientists are able to show that race does not accurately capture human genetic diversity, yet, at the same time, others claim that it is a useful proxy to best capture human genetic diversity, a proxy that is especially useful in biomedical settings.

The upsurge of claims that race is, in fact, a useful taxonomic concept for humans seems to be driven by several factors. First, genomic technology has enhanced our ability to examine the 0.1 percent of nucleic acids in the human genome that, on average, vary between individuals. Some scientists are relying on the race concept to make sense of the genetic variation in this small sliver of our genomes. Second, the critical task of understanding and reducing known disparities in health—including disparities in disease including heart disease, cancers, and diabetes—has researchers looking at all possible explanations, including genetic ones, for disparities in health outcomes. Fueled by programs including the National Institutes of Health's "Healthy People 2010" and the Centers for Disease Control and Prevention's "Racial and Ethnic Approaches to Community Health," the search for the underlying causes of these disparities is a national healthcare priority. The renewed focus on race and genetics suggests, however, that an analysis of the complex relationship between individuals, populations, and health will be surrendered to a simplistic, racialized worldview. An inability to digest (and frequently even to acknowledge) these complexities restricts scientific theory and practice to simplicity when complexity is needed. This underlies the drive to correlate race, genetics, and health disparities.

Third, and finally, the history of the biological race concept suggests that race is deeply embedded in scientific and social thought, and that racialized thinking was an integral part of genetics in the twentieth century.[38] This history has shaped scientific thinking about human difference, as well as popular thinking about that difference.[39] From W. E. B. Du Bois's attack on the biological race concept to Theodosius Dobzhansky's reconsideration of the race concept in the context of modern biology to Richard Lewontin's molecular analyses that showed the shortcomings even of a population-genetics- and evolutionary-biology-based concept of race, debates about the concept of race in American scientific thought remain largely unchanged. By this I do not mean to suggest that the race concept itself has not changed. Indeed, this chapter documents its evolution. Instead, we seem to be having very similar arguments about race and about human difference at the dawn of the twenty-first century as we were at the outset of the twentieth. On one side, advocates of the race concept claim technological and methodological improvements allow them to examine human diversity with increasing precision that is disconnected from any social prejudices about human difference. On the other side, critics of the race concept insist that it is a flawed, inaccurate way to measure human genetic diversity that is inseparable from social prejudices about human difference and that biologists need to develop other methodologies to measure genetic differences between human populations. Furthermore, for those scientists currently seeking to redefine the meanings and usage of race in examining health disparities, this history suggests that they are falling into the same trap that Dobzhansky fell into half a century ago; a belief that despite racism and prejudice the race concept could be a "device used to make diversity intelligible and manageable."[40] History has shown that even acknowledging that race has both a social and a scientific meaning cannot disconnect the concept from its typological and racist past (or present). Despite the best intentions of many scientists and scholars, race will always remain what Ashley Montagu once called a "trigger word; utter it and a whole series of emotionally conditioned responses follow."[41]

We are a genetically diverse species, and there is meaning in that diversity. But we as a species seem thus far unable to reliably distinguish between the scientific ramifications and the social meanings of human difference. Race is an historical, not a scientific, term. Yet, until the scourge of racism is eliminated from our lives and institutions, developing methods unburdened by racial ideology to study human difference will be an impossibility. Biology may develop new ways of studying human populations that to whatever degree distance themselves from the biological race concept and its historical baggage, and those new methods may, in fact, be an improvement over where we stand now. But that social and natural scientists have been rejecting, abandoning, and discrediting the race concept for over a century suggests that for now the race concept is here to stay.

NOTES

1. See, for example, P. C. Ng, Q. Zhao, S. Levy et al., "Individual Genomes Instead of Race for Personalized Medicine," *Clinical Pharmacology and Therapeutics* 84

(September 2008): 306–9; Svante Paabo, "The Mosaic That Is Our Genome," *Nature* 421 (2003): 409–12; Marcus Feldman, Richard Lewontin, and Mary-Claire King, "Race: A Genetic Melting-Pot," *Nature* 424 (2003): 374; Alan Goodman, "Why Don't Genes Count (For Racial Differences in Health)," *American Journal of Public Health* 90 (2000): 1699–72.

2. Histories of the race concept include: Bruce R. Dain, *A Hideous Monster Of The Mind: American Race Theory In The Early Republic* (Cambridge, MA: Harvard University Press, 2002); Audrey Smedley, *Race In North America: Origin And Evolution of a Worldview* (Boulder, CO: Westview Press, 1999); William Stanton, *The Leopard's Spots: Scientific Attitudes Toward Race In America, 1815–59* (Chicago: University Of Chicago Press, 1960); Nancy Stepan, *The Idea Of Race In Science: Great Britain, 1800–1960* (Hamden, CT: Archon Books, 1982); Elazar Barkan, *Retreat Of Scientific Racism: Changing Concepts Of Race In Britain And The United States Between The World Wars* (New York: Cambridge University Press, 1992); Stephen Jay Gould, *The Mismeasure Of Man* (New York: Norton, 1996).
3. Thomas Jefferson, *Notes on the State of Virginia* (Chapel Hill: University of North Carolina Press, 1955), 138–40, at 143.
4. Recent DNA evidence suggests, for example, that Jefferson's relationship with his slave Sally Hemings produced at least one child. See E. A. Foster et al., "Jefferson Fathered Slave's Last Child," *Nature* 396 (1998): 27–28; Eric S. Lander and Joseph J. Ellis, "Founding Father," *Nature* 396 (1998): 13–14.
5. Frank M. Snowden, *Before Color Prejudice: The Ancient View of Blacks* (Cambridge, MA: Harvard University Press, 1983), 107.
6. Ivan Hannaford, *Race: The History of an Idea in the West* (Washington, DC: Woodrow Wilson Center Press, 1996), 14, 17–60.
7. George M. Fredrickson, *Racism: A Short History* (Princeton, NJ: Princeton University Press, 2002), 17
8. Hannaford, *Race*, 123.
9. Georges Louis LeClerc, *Natural History: General and Particular* (London: T. Cadell and W. Davies, 1812).
10. Agnes Smedley, *Race in North America: Origin and Evolution of a Worldview* (Boulder, CO: Westview Press, 1999), 164.
11. Gould, *Mismeasure of Man*, 403, 408.
12. Drew Gilpen Faust, *The Ideology of Slavery: Proslavery Thought in the Antebellum South, 1830–1860* (Baton Rouge: Louisiana State University Press, 1981), 14; Gould, *Mismeasure of Man*, 70; Jason E. Lewis, David DeGusta, Marc R. Meyer, et al., "The Mismeasure of Science: Stephen Jay Gould versus Samuel George Morton on Skulls and Bias," *PLos Biology* 9(2011): e1001071.
13. Francis Galton, *Inquires Into Human Faculty and Its Development* (New York: MacMillan, 1892), 25.
14. Celeste Condit, *The Meanings of the Gene* (Madison: University of Wisconsin Press, 1999), 27.
15. William Provine, "Genetics and Race," *American Zoologist* 26 (1986): 857–87.

16. Charles B. Davenport, "Do Races Differ in Mental Capacity?" (lecture, March 1928), Charles B. Davenport Papers, American Philosophical Society Library. This research was published in Charles B. Davenport and Morris Steggerda, *Race Crossing in Jamaica* (Washington, DC: Carnegie Institution of Washington, 1929).
17. Paul Popenoe and Roswell Hill Johnson, *Applied Eugenics* (New York: The Macmillan Company, 1933), 283–84.
18. Gregory Dorr, *Segregation's Science: Eugenics and Society in Virginia* (Charlottesville: University of Virginia Press, 2008); Edwin Black, *War Against the Weak: Eugenics and America's Campaign to Create a Master Race* (New York: Four Walls Eight Windows, 2003), 159–82.
19. Kelly Miller, *Race Adjustment: Essays on the Negro in America* (New York: Neale, 1909), 44.
20. *The Health and Physique of the Negro American*, ed. W. E. B. Du Bois, vol. 11, *The American Negro: His History and Literature* (New York: Arno Press, 1968), 16.
21. Theodosius Dobzhansky, *Mankind Evolving: The Evolution of the Human Species* (New Haven, CT: Yale University Press, 1962), 263–66.
22. Theodosius Dobzhansky and L. C. Dunn, *Heredity, Race, and Society* (New York: Penguin Books, 1947); Ruth Benedict, *Race: Science and Politics* (New York: Modern Age Books, 1940); Ralph Bunche, *A World View of Race* (Washington, DC: The Associates in Negro Folk Education, 1936); Jacques Barzun, *Race: A Study in Modern Superstition* (New York: Harcourt, Brace & Company, 1937).
23. Theodosius Dobzhansky, *Genetics and the Origin of Species* (New York: Columbia University Press, 1982), 60.
24. Gunnar Myrdal, *An American Dilemma: The Negro Problem and American Democracy* (New Brunswick, NJ: Transaction Publishers, 1996), 15.
25. Ashley Montagu, *Statement on Race: An Annotated Elaboration and Exposition of the Four Statements on Race Issued by the United Nations Educational, Scientific, and Cultural Organization* (New York: Oxford University Press, 1972), 7–13, 65.
26. L. C. Dunn and R. A. Fisher to Métraux, October 3, 1951. Leslie Clarence Dunn Papers, American Philosophical Society Library.
27. John P. Jackson, Jr., *Science for Segregation: Race, Law, and the Case Against Brown V. Board of Education* (New York: New York University Press, 2005).
28. Thomas Borstelmann, *The Cold War and the Color Line: American Race Relations in the Global Arena* (Cambridge, MA: Harvard University Press, 2001), 55.
29. Richard Lewontin, "The Apportionment of Human Diversity," *Evolutionary Biology* 6 (1972): 381–98; Richard Lewontin, *Human Diversity* (New York: W. H. Freeman, 1982).
30. Guido Barbujani, Arianna Magagni, Eric Minch, and L. Luca Cavalli-Sforza, "An Apportionment of Human DNA Diversity," *Proceedings of the National Academy of Sciences, USA* 94 (April 29, 1997): 4516–19.
31. Arthur Jensen, "How Much Can We Boost IQ and Scholastic Achievement?" *Harvard Educational Review* 39 (1969): 1–123.
32. Harold M. Schmeck, Jr., Nobel Winner Urges Research on Racial Heredity," *New York Times*, October 18, 1966.

33. Henry Louis Gates, Jr., "Critical Remarks," in *Anatomy of Racism*, ed. David Theo Goldberg (Minneapolis: University of Minnesota Press, 1990), 326.
34. Rick Weiss and Justin Gillis, "Teams Finish Mapping Human DNA," *Washington Post*, June 27, 2000.
35. F. S. Collins and M. K. Mansoura, "The Human Genome Project: Revealing the Shared Inheritance of All Humankind," *Cancer* 92 (2001): S221–S225.
36. N. Risch, E. Burchard, E. Ziv et al., Categorization of Humans in Biomedical Research: Genes, Race, and Disease," *Genome Biology* 3 (2002): 1–12.
37. Robin M. Henig, "The Genome in Black and White (and Gray)," *New York Times Magazine*, October 10, 2004, 47.
38. William Provine, "Genetics and Race," *American Zoologist* 26 (1986): 857–87.
39. Vanessa N. Gamble, "A Legacy of Distrust: African Americans and Medical Research," *American Journal of Preventive Medicine* 9 (1993): S35–S38.
40. Dobzhansky, *Mankind Evolving*, 262–66.
41. Montagu, *Statement on Race*, 65.

2

NATURAL SELECTION, THE HUMAN GENOME, AND THE IDEA OF RACE

Robert Pollack

This chapter discusses the history of humanity as a single species, born of an ancestral species some hundreds of thousands of years ago in Africa. The history of our single species tells us that all people who were ever born anywhere on Earth have been, are, and will be descendents of Africans. Because all human beings are members of one species, all concepts of "race" that place one set of humans aside as in some way more or less fit or worthy than another set, must be in conflict with the facts of nature. The persistence of imaginary, false notions such as "race" is an example of the most remarkable characteristic of all members of our species: our imaginations. The emergence in our species of brains capable of mental worlds and self-awareness has paradoxically produced both the science that reveals these facts of our history and our biology, and the dreams of perfection that keep such imaginary notions as "race" and racism alive despite these facts. The chapter concludes with the optimistic observation that the same DNA-encoded brains that can have any thought are also therefore capable of learning these facts from science and choosing to discard the fantasies of "race."

THE IDEA OF RACE

Can there be a plausible biological basis for the negative category called race? As we use the term in America today, "race" is an idea of a particular sort. The idea of "race" uses biological differences, but it is not about biological differences. It is the classic example of the idea of a negative category, one that defies definition because it lacks content. Race is not what a racist may say it is; it is simply whatever a racist thinks he or she is *not*. What can knowledge of the facts of natural selection and the human genome contribute to our understanding of such an idea?

To begin at the beginning, such knowledge can give us a sense of the radical novelty of humanity as a single species in the context of the long histories of the universe, and of life on our planet. Unlikely and exotic as it may seem, the best explanation for the facts of astronomy today is that the universe, all space and time itself, began at a point, all at once, some 13.7 billion years ago.[1] As space has expanded from that point to beyond the furthest galaxies, it has never once been separated from the passage of time, with one exception. Time flows neither in our imaginations nor our dreams as it does in nature; we may imagine timeless idealizations, perfectible futures, and heroic pasts all at once.

Recently, in terms of the history of the universe—about 4 billion years ago—something as improbable as anything we might imagine occurred here on the Earth. In the salty seas, and apparently initially at random, clusters of atoms found in interstellar space got hooked up into long strings, and a very rare sequence of those strung-out clusters acquired the capacity to make a copy of itself, preserving the sequence of the subunits in the string. A self-replicating string preserves the information in the sequence of that string, so long as the copies themselves can make more copies of themselves in turn. One of these self-copying strings of chemical letters, DNA, has been copying itself ever since.

DNA is a chemical of great informational density, a text of great importance. As far as we know today, it is a new thing in the history of the universe, having appeared on our planet and, so far, nowhere else we know of. Of course, self-copying by itself does not explain why life emerged on our planet; it is necessary, but not sufficient. The second requirement for life is that the different strings of subunits of the self-copying DNA carry meanings, and that one of these meanings be the capacity to assist the DNA in making more copies of itself. Thereafter, any version of DNA encoding a novel strategy for the survival of DNA after copying would itself be preserved as the novel meaning of that new sequence of DNA. This second step, Darwin called it natural selection, is both necessary and sufficient to explain the history of life on Earth, from the first DNA-encoded organisms, to us and all the species of creatures and plants alive on Earth today.

Our form of life, which emerged out of the same process of natural selection from DNA variations that has operated to produce the living novelties of this world since then, has been around only a very short amount of time indeed.

Think of each million years since the beginning of the universe as a page in a book. Today that bookshelf of the Universe would hold 30 volumes of 450 pages each (see fig. 2.1). The first 21 volumes would have nothing in them about life. Both DNA sequence and fossil evidence agree that the informational molecule DNA would have been born some time in volume 21, because archeobacteria, the first forms of life, would appear in the seas in volume 22.

Bacteria would continue to be the only shape life took for volumes 23 and 24 as well, though the ones emerging in volume 24 would change the planet's atmosphere to one rich in oxygen by bacterial photosynthesis. Big-celled forms of life like paramecia and diatoms would appear for the first time in volume 25. Living things made of many big cells would appear in volume 27. Animals would remain in the seas where life had begun until the first forms of animal life that appeared on land, the first tetrapods, march on shore at the end of volume 29.

Dinosaurs would appear in the middle of volume 30. They would for the most part be wiped out by an asteroid on page 385. Only the last 65 pages of the last volume would have anything to say of significance about mammals like the cat. The last ancestor of both us and our nearest living relative, the chimp, would have lived and died only by page 443 of the most recent volume, 17 million years ago. From that ancestor many other ancestral hominoid species would follow, each coming and going in the last 10 pages.

On the last tenth of the last page of that last volume humans would have a note about our emergence in Africa. And then, somewhere toward

FIGURE 2.1 A. Pollack and R. Pollack, "We Have Been Around Only a Very Short Time," *CrossCurrrents*, Spring 2007, 136.

the last sentence would be the emergence of language, texts, and, in that *mental* world, thoughts of imagined and imaginary creatures like Alice in Wonderland. The period at the end of that sentence would hold the time since science emerged in our mental worlds as a social activity with the capacity to understand all this.

And so at last we come to "race" and racism. In this last eye blink of universal timekeeping, we find ourselves entranced by two notions that share the same persistence in our minds and the same imaginary quality as Alice herself: first that a person is no more than what that person has inherited in her DNA, and second that a person's race is merely the clearest example of that presumption. The first is a dream because the facts of science assure that our mental worlds are *not* encoded in our DNAs. Any brain can imagine, learn, teach, remember, or forget any idea, regardless of the ancestry of the person whose mind is emergent in that brain, and regardless of whether that idea does or does not reflect the facts of nature. Perhaps the most self-serving and punitive example of such a dreamt idea is the notion that "genes are destiny." They are not, and the dream that they are, must certainly prevent a person from celebrating the freedom to think a new thought, which is in fact the birthright of every human brain and mind.

We hold to these dreams even though for many decades science has been able to establish, through the repeated failure of all experimental attempts to disprove it, that in fact the contrary is the case. Any person's genome—his or her complement of two copies of each of about ten thousand genes, one copy of each from each parent—is no more the complete statement of that person's life and character, than any single version of a canonical text is the complete statement of a living religion. Everything interesting in both cases is the product of interpretation and social interaction. Appearances, as well as more subtle aspects of a person's individuality, begin with the information encoded in DNA, but anyone who knows an identical twin also knows that person to be unique, despite the presence of another person with the same DNA text in each cell.

From any one person to another, unrelated person, the differences in base-pair sequence—letters in the text—for the coding region of any gene studied come to about one in a thousand. The DNA sequences of the cells of two siblings are more closely related, but not that much more so, as each of the two copies of each gene have only a one-in-four chance of being the same in each child. Cumulatively over all ten thousand genes, even brothers and sisters are different in DNA sequence to almost the same degree as strangers. That is very different indeed: with three billion letters in the DNA we inherit from each of our parents, one base-pair in a thousand comes to three million sequence differences between any two people.

Beyond that fact, we should remember that the number of our ancestors doubles with every generation, so that in the past few centuries—the past ten generations—each of us has had more than a thousand ancestors, and all of them would have had this same larger number of DNA sequence differences from each other. Imagine a canonical text with that many variations from century to century and from copy to copy: no chance of any version possibly being the only true one. Yet that is precisely the meaning of the otherwise empty notion of "race."

A racist holds that one person and all her ancestors can be the product of a set of DNA sequences so restricted in its variation that no one else but other people who look like that person have ever had or will ever have any of those sequences. The data are in, and this is not possible. It is not merely that we do not have such data; we do have enough data to be sure that the notion of oneself as the member of any such genetically restricted group, let alone one that is defined by a singular closeness to any imagined ideal, is merely a fantasy.

All that makes our genomes human, and all that makes us human in a biological sense, is that these six billion different genomes, and only they, are all capable of coming together with each other through sperm and egg to make another generation of people. The sieve of natural selection assures us that no matter what the differences in our DNA sequences, all of us are here because our ancestors' DNAs contained the capacity to encode the structures for fertile reproduction in their bodies. Everything else encoded by their DNA, and therefore by ours, that makes us different from one another, is either in service to that fact of the necessity for fertility so that the species and its DNA will survive, or it is a difference carried along by that fertility because it does not get in the way of it.

No rational explanation of what it is to be human, then, can possibly begin with the claim that one set of these exceedingly large number of genetic variants encoding fertile but different people encodes a fully human person, and another only apparently human but not worth the full recognition and rights of "one of us." The biology is clear: there is no chance of some human genomes being special and others not; the biology of us makes us truly all equal.

The presumption of "race" in the American context runs up against a second fact about our history as a single species. Our species is African in origin; we are all the very recent descendants of Africans. The evidence for this comes from many quarters, but in our terms the DNA evidence is most interesting. Because Africa is the home of us all, people today who are the descendents of the original people—hundreds of millions of Africans—have the greatest genetic diversity of all human subpopulations. This is because those subpopulations who left Africa to cover the other continents left close relatives behind, and their descendents are the people who live in central Africa today. Of course the DNA sequences of people everywhere are also in flux as new DNA changes (mutations) pass the test of natural selection and persist, whether by being of no consequence or—rarely—by being advantageous for the survival of offspring. Still, the life of any species is measured in millions of years. We are therefore very young, and the descent of all of today's many "ethnicities" and "races" from people who lived in Africa only tens of thousands of years ago is well established.

America is also a genetically diverse place, but for a different reason. We are, in the words of President John F. Kennedy, "a nation of immigrants."[2] Setting aside most "African Americans" for a moment, the rest of the diverse American population is a set of immigrant subpopulations, each the descendents of one of the initial emigrant subpopulations to leave Africa tens of millennia ago. In that sense, though only some of us are African Americans, as a nation of immigrants from all over the globe, we are still all American Africans.

The many different versions of a stretch of chromosomal human DNA still found today in East African populations have been studied in other populations as markers of first African subpopulations to have left the ancestral homeland for the one or another of the lands reached by a series of migrations that began no later than sixty thousand years ago. These DNA fragments confirm archeological evidence that the most recent human migration to arrive at its final destination was the one that settled at the southern tip of the Americas about ten thousand years ago.[3]

The first people we would recognize as our ancestors if we met them today were Africans whose skins were dark. From these first ancestors, emigrants migrated throughout Africa and then to the Middle East, Europe, Asia, Oceania, Russia, and North America, finally ending at the southern tip of South America ten thousand years ago. Once the planet had been colonized by African emigrants, the itch to move on did not go away, nor has it even today. There is a difference between those first migrations and later immigrations and invasions. For the last ten thousand years, no migration would have been likely to settle in territory not already occupied in part by descendents of that initial migration. The resulting wars and conquests form the narrative of what we are pleased to call modern civilization.

Our way of understanding our history as a nation needs revision in light of what we now understand is the pattern of migration of human populations.[4] Universal African patrimony makes the insult of American racism more stupid, but not less dangerous, than any other dehumanization. The series of European "discoveries" of the Americas in the past millennium were simply secondary migrations to, and conquests of, the lands first occupied by the original African settlers of North and South America by descendents of the original African settlers of Europe. That some of these "conquerors" first entrapped then enslaved Africans of their day, that these Africans arrived on the shores of the Americas in that fashion, and that the founders of the United States then enshrined the legal nonpersonhood of their descendents in the very first article of the country's Constitution makes this compelled migration only more poignant and ironic: "Representatives and direct Taxes shall be apportioned among the several States which may be included within this Union, according to their respective Numbers, which shall be determined by adding to the whole Number of free Persons, including those bound to Service for a Term of Years, and excluding Indians not taxed, *three fifths of all other Persons.* (my emphasis)."[5]

"Indians not taxed" were at that time members of unsubjugated nations. However inconvenient their presence was to notions of the manifest destiny of Europeans on this continent, they were powerful and free enough to be understood to be people in their own right. "Other persons" were not people. People can either be ignored or taxed and given the vote; "other persons" could not vote, could not become voters, had no rights, and could be bought and sold. But they had political reality and political utility. For the purposes of counting the number of seats in the US House of Representatives, "other persons" would be counted, so that a slave owner who owned one hundred "other persons" would be counted for that purpose as if he were sixty-one voters.

The recent election of the United States' first President of acknowledged and recent African ancestry has not closed this sorry history, but it has transformed its irony into simple failure. We have no national monument to "other persons" per se, while in the past few years we have seen a National Institutes of Health (NIH) initiative to examine human DNA for evidence of race. This NIH project intends to find versions of genes that are in everybody of one race, or "ancestry" (a euphemism for race in this context), but which are never found in the genomes of people not of that race, so that the complexity of a real person, with all her uniqueness of character, history, and potential for change, may be reduced to the presence or absence of such a DNA sequence. Enough is known of human genetic diversity to make this an unlikely outcome in any event. Of all such putative DNA differences, it would be irrelevant at best and racist at worst to seek to use DNA differences associated with skin color differences. DNA differences responsible for skin color differences, the gold standard of American racial classification, turn out to be subject to very strong and rapid natural selection.

Here is how one scholar of the evolution of human skin and skin color put it in a recent review: "Dark skin evolved *pari passu* with the loss of body hair and was the original state for the genus *Homo*. Melanin pigmentation is adaptive and has been maintained by natural selection. Because of its evolutionary lability, skin color phenotype is useless as a unique marker of genetic identity."[6] Our shared African ancestors were dark-skinned because our species had emerged from hairless variants of an ancestral species, and naked apes like us were most likely to survive under the UV-rich rays of a tropical sun with the pigment melanin robustly produced by cells under their skin. As our African ancestors migrated away from the equator to more northern latitudes, the sun's rays were no longer so much of a selective agent, and lighter-skinned variants of human DNA conferred the advantage of permitting enough UV light to reach the blood under the skin, so that a person would be less likely to suffer the consequences of a Vitamin D deficiency. When these light-skinned early Europeans and Asians returned by further migration to the equatorial regions of Asia and the Pacific Islands, their descendents once again emerged as dark-skinned.

In sum, DNA samples from an individual cannot be used for any purpose related to the skin-pigmentation notion of "race," because the DNA differences associated with pigmentation will reflect the range of skin colors of one's most recent ancestors. Worse, when the categories of a "race" are attached to the DNA differences responsible for the intensity of melanin production, the result will be a biologically useless but politically powerful justification for the presumption that the DNA sequence signaling dark skin is also a signal for any data-free racist presumption of what a person will necessarily be when this DNA says he or she is "not like one of us."

THE BIOLOGY OF AN IDEA

To the extent that race is a negative category, racial differences cannot be the result of any number of genetic differences at all. This is the origin of the arbitrary nature of any and all racial categories and the absurdity of the recent evasive slide in America from "racist race" to politically correct "identity politics." In any event, the negative category of "whom I am not" can never be reducible to a countable number of genetic differences. But beyond that, there is the question of purpose. Even looking for genomic data on this question will not be of interest to the racist, who says or thinks, "if you are not like me, you can be anything else at all, I don't care." Why, then, should it be of interest to anyone else? As a negative category, "race" is an idea that resists scientific elaboration, but it is a powerful idea nevertheless, and as such it needs to be studied and understood. Where, then, do ideas come from?

There are about three billion letters in the human genome. But there are about a millionfold more synaptic connections in a human brain at birth than there are letters in any human cell's canonical text. These synaptic connections, the basis of all mental activity later in life, cannot have all been specifically encoded by our genomes. At birth some are not functional, nor are many stable or specific; synaptic connections harden into circuits only later. We begin with a tissue that becomes a mind by social interaction. Our DNA encodes, in other words, a "learning machine." The learning machine is very complicated: it requires that most human genes be present in functional versions; that is, half or more of the genes in the genome are active in the nervous system, and for the most part only in the brain.

What these genes encode is the capacity of synaptic connections to be stabilized by use through the activation and repression of genes in nerve cells. The learning machine starts up at birth at the latest, activated by the initial input signals from the organs of perception. This is the mechanism by which the mind slowly emerges from the brain, through imitation of the minds of those people with whom the infant interacts. The experiences of the first two years, before language, lay down much of the stable circuitry of the thinking brain. But even after our formative years, the mature brain forever retains plasticity in its circuits, and it never loses the capacity to link past with present experience by familiarity of synaptic pattern. Synaptic connections are made and broken throughout life; these are experienced variously as sensation, perception, memory, repression, and, most important, one's ongoing teaching and learning.

Make no mistake, this may be a difficult idea to take in, but, in terms of biology, it is the best we can say about how we understand ourselves. Whenever in your entire life you learn, or forget, or remember, or deeply feel any emotion, these mental events are not merely accompanied by the reorganization of cell-cell circuits within your brain: they *are* those reorganizations. Reading these words with comprehension is such a reorganization, remembering them is another, telling someone you learned something odd is still another rewiring of your brain. Because we never stop thinking and feeling, we now know that in ways we do not yet fully grasp nor measure our brains never stop this ceaseless reweaving of their circuitry.

The "learning machine" that is the human brain cannot weave and reweave itself in isolation. Mental development and mental health both require adequate social interaction from birth on; absent that, sociopathic disasters ensue. Racism is one of these disasters. Whoever is cast as the "Other" by adults when they interact with their children will become the "Other" to those children. When "race" is learned in this way, it is a biological event, in that the synaptic wiring of associations in the brain of the child will have mimicked those in the brain of the adult. This form of inheritance is not through DNA, but it can be as stable, and as long-lasting, as genetic inheritance. But we must be clear: it is social, not genetic.

As a social construct, "race" may appear as a useful idea when it is couched in contexts of service, in particular when it is used as a way to identify an American population that might specifically benefit from a medical or other social intervention. But that apparent utility will always be reduced to close to nil by the complexity of our history and the diversity of the human genome. We have seen a drug marketed to "African American" Americans on the argument that the drug requires the presence of a single DNA sequence found in many Africans and not many Europeans (see chapters 7 and 8). We know already that this allocation scheme has assured that the drug has been wasted on some number of Americans with both European and African ancestors who are identified as "African American" but who did not inherit that version of that DNA, while a number of "white" Americans of similar parentage who did inherit it and therefore could benefit from the drug do not get to use it.[7]

Surely if the DNA sequence that is required for drug action is known, then a simple assay for that sequence, not a measure of "race," is the appropriate prognostic test. The same argument holds for the high frequency of disease-associated DNA sequences in populations descended from a small number of ancestral families, like the Ashkenazi Jews of Eastern Europe. These are sometimes called "Jewish Diseases." In fact, the ancestry of Eastern European Jews includes their almost complete obliteration about 350 years ago; all Ashkenazi Jews today are the descendents of a small number of surviving families, so these should probably be called "survivor diseases" instead. Whenever DNA assays for these DNA differences are done as broadly as they should be, many people who have that DNA sequence and therefore may benefit from the knowledge are not Jewish, but turn out by that assay to learn of the likelihood of having a forgotten or denied Jewish ancestor.

Here, finally, is an example of how rapidly and completely the idea of "race" may be understood in two different ways, depending on which set of presumptions informs our mind when we learn that information. The "Human Diversity Project" has examined thousands of peoples' DNA, sampled from all over the planet, and sorted short DNA sequences that are most common in people from one or another continent today. People present a DNA sample to this organization and are told they are "25 percent African and 75 percent European," or the like.

How do we understand this? On the one hand, we may say that since we are all Africans initially, this must mean that among the ancestors of that person are some people who left for Europe about forty thousand years ago and others who never left. Or, we could say that this person has a one-quarter dose of "Other" and is therefore, in the old pre-DNA language, a "quadroon." I would argue that until the notion of the "quadroon" has entirely left our culture—and it has certainly not—we should choose to be very careful not to use any language that speaks of a person in percentages of anything, let alone ancestry or "race." The racist thinks of everyone in the "Other" category as if they were genetically identical clones: "all you people look alike to me." The irony of thinking of the "Other" this way is more perfect in the American case than any other. Here the "Other" is likely to be a descendant of Africans, who are today the most genetically diverse of all people. It is the racists who, thinking alike despite all facts, form a clone, not a genetic clone, but a social one. That is why there is no contradiction between thinking of "race" as a social construct, the product of racism, and thinking about it in the language of genetics.

The best example of the ease with which a decent and insightful person may fall into the trap of imagining some members of our robust, outbred species to be not merely the "Other" but to actually be another species entirely is found not in the atrocities of the last century, but in the ruminations of the brightest mind of the century before. Charles Darwin, whose bicentennial we celebrated in 2009, followed his momentous work, *On the Origin of Species*, with another book, *The Descent of Man*, intended to deal with the many questions about humanity's origins and nature raised by the insights of his first book.

Of these questions, the most fraught with risk was the matter of whether or not all humanity was one species. He weighed the evidence, explained the difficulties, and concluded that the main observational impediment to the notion that "humanity" is an agglomeration of different species was the disconcerting fertility of any and all matings among people of any supposed racial "species." The cornerstone of his first book was the notion that life persists through time only in species, whose boundaries he operationally defined by the failure of individuals from different species to produce fertile offspring.

Unexpectedly, and to my eye disconcertingly, he concluded by overturning his own insight to protect the notion that the "Other" and he were not of the same species:

> Man in many respects may be compared with those animals which have long been domesticated, and a large body of evidence can be advanced in favour of the Pallasian doctrine, that domestication tends to eliminate the sterility which is so general a result of the crossing of species in a state of nature. From these several considerations, it may be justly urged that the perfect fertility of the intercrossed races of man, if established, would not absolutely preclude us from ranking them as distinct species.

To speculate on the motives of a great thinker when he or she trips up in this way seems to me unseemly, but Darwin's contradiction of his own findings may reflect the fact that, despite his insight, he was a product of his time and culture, and that his brain had been woven and rewoven in its circuitry so that he found himself, in the absence of evidence, comfortable to think in a racist manner. In any event we no longer have Darwin's freedom to ruminate along these lines. We know that we are a single species at every level except in our imaginations, and we need not look back even a century to see the horrendous damage done by the counternotion, when it emerges as the policy of a state.

CONCLUSION

Human developmental biology and genetics must be present in any serious future discussion of "race" as an idea. But because "race" is an idea of great toxicity, the proper genetics here is not the genetics of skin, or of lips or hair or the shape of one's rump (an embarrassment even to write here), but the genetics of human neonatal neural development. That is where we will find the biology of the ineluctable and irreducible, but *reversible*, role of parental modeling in the emergence of a mind and of ideas in that mind. It is through the constant reweaving of those circuits in social interaction and in isolation that we experience the twin gifts of natural selection: free will and the chance to change our minds.

NOTES

1. Dennis Overbye, "With Updated Hubble Telescope, Reaching Farther Backing Time," *New York Times*, January 12, 2010.
2. John F. Kennedy, *Nation of Immigrants* (New York: Harper and Row, 1964).
3. See G. Stix, "Traces of a Distant Past," *Scientific American*, July 2008, 56.
4. US Constitution, article 1, clause 3. See also Stix, "Traces of a Distant Past," 56.
5. N. Jablonski, "The Evolution of Skin and Skin Color," *Annual Review of Anthropology* 33 (2004): 585.
6. www.fda.gov/bbs/topics/news/2005/new01190.html.
7. Charles Darwin, "On the Races of Man," in *The Descent of Man* (New York: Appleton, 1880), 172.

PART II

FORENSIC DNA DATABASES, RACE, AND THE CRIMINAL JUSTICE SYSTEM

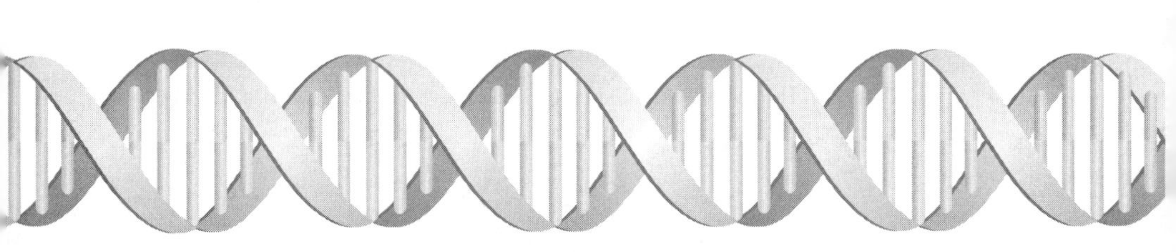

3

RACIAL DISPARITIES IN DATABANKING OF DNA PROFILES

Michael T. Risher

Of the hundreds of thousands of arrests every year in California on suspicion of a felony, more than 325,000 in 2008, approximately 30 percent never lead to any conviction.[1] In the US justice system, people who are arrested but never convicted are presumed innocent. A disproportionate number of these innocent arrestees are people of color. As of January 1, 2009, these people have been forced to let the State of California take a DNA sample, analyze it, and include the resulting profile in a criminal database, to be compared evermore with crime-scene evidence. As discussed below, although there are procedures for some of these people to try to have these samples and profiles expunged, they may require that people wait three or more years before even requesting expungement and will require that people seek the help of a lawyer to navigate the procedures, which are not even available to people who have ever been convicted of a felony. As a result, the overwhelming majority of people arrested but not convicted of any crime are unlikely even to try to get their samples destroyed. Tens of thousands of profiles taken from innocent people will thus remain in these criminal databases. The consequence will be a magnification of the current racial disparities—at least in absolute terms—in our criminal justice system, as more and more people of color's DNA profiles are included in databases that make them potential suspects whenever DNA is recovered from a crime scene.

The effects of this disproportionate inclusion of people of color in the databanks are made clear by the other papers in this series on genetics and race. What is perhaps less clear is how our criminal justice system, which promises equal justice under the law, can tolerate this injustice. After outlining the legal growth and transformation of DNA databanks, this chapter examines how various steps in our criminal justice system create and magnify racial disparities, and explores how the law makes it nearly impossible to effectively address the problem. It also looks at how taking DNA samples at various stages in this process may affect these disparities and the factors that cause them. I use as my primary example California's system because it is one of the world's largest criminal justice systems in one of the nation's most diverse states. It is also the system in which I have practiced law for the last decade, and is representative of where DNA databanks throughout the country will likely be in the next few years as more and more states and the federal government collect DNA from arrestees.[2]

A VERY BRIEF INTRODUCTION TO DNA DATABANKS

DNA databanks comprise two distinct components: the actual biological samples and the computerized database of the profiles generated by analyzing these samples. In criminal-justice databanks, the biological samples are collected from crime scenes (forensic samples) and from known individuals (known samples). Until recently, known samples were usually obtained by drawing blood, although now most states and the federal government primarily obtain samples by swabbing the inside of the person's cheek to collect skin cells.[3]

The government analyzes both forensic samples and known samples to create DNA profiles, which are essentially a digitized description of twenty-six parts of the person's nuclear DNA molecule. The profiles are then uploaded to the Combined DNA Index System (CODIS), a centralized, searchable law enforcement database accessible to state, federal, and international law enforcement agencies.[4] CODIS was created by the FBI in 1994 after Congress authorized it to establish a national DNA database to link existing state and local databanks.[5] The biological samples themselves are retained by the local police or crime lab for later testing.

Once an arrestee's profile is uploaded into CODIS, it is compared to the thousands of crime-scene samples in the CODIS forensic database. As long as the arrestee's profile remains in CODIS, any new crime-scene samples will be searched against it. When an arrestee profile exactly matches a crime-scene profile, CODIS automatically notifies the agencies that provided the samples. Then that agency will usually provide the identity of the arrestee to the police authority with jurisdiction over the crime so that the latter can follow up.[6]

THE GROWTH OF DNA DATABANKS

DNA databanks have grown exponentially in the last decade as new laws have expanded the range of people subject to having their DNA forcibly seized, analyzed, and the resulting profile databanked. California's databank is a good example of this. It was originally conceived as a way to connect people convicted of serious violent crimes with other such crimes in which DNA evidence is most useful. Thus, the original 1989 DNA-collection law established a databank and required people convicted of murder or a felony sex offense to provide DNA samples before they were released from custody.[7] The state department of justice had the authority to analyze these samples and include the resulting analysis in the new statewide databank.[8] From today's perspective, this program seems quite limited: the only people subject to having their DNA databanked had been convicted of very serious crimes, either by pleading guilty or after the charges had been proved beyond all reasonable doubt to a jury.

Because conviction of a serious felony has long resulted in serious consequences, for example, incarceration, followed by a period of government supervision and surveillance during probation or parole, the additional intrusion of having to give a DNA sample was unlikely to be the most serious consequence from the conviction. On the other side of the equation, the benefits of requiring people convicted of serious crimes to provide DNA samples were substantial: society has a strong interest in solving serious violent crimes, and people who had been convicted of such offenses are statistically more likely than others to commit other violent crimes. In addition, crimes of violence that may involve struggles, including murder and sexual assault, are more likely than most to involve DNA evidence from semen, blood, or other bodily tissue.

But the law soon began to expand to include more people. The first steps were modest: in the late 1990s, new crimes were added to the list of qualifying offenses, and the law was amended to require that samples be taken immediately after conviction, rather than just before release. The latter change was enacted as the focus shifted from preventing new crimes to solving old crimes.[9] In 2004, California voters enacted Proposition 69, drastically expanding the database. The two biggest changes were that, as of November 2004, every person convicted of *any* felony—which can include simple drug possession, shoplifting, or even intentionally writing a check without sufficient funds to cover it[10]—has had to provide DNA samples. And, as of January 1, 2009, every person *arrested* for a felony in California must give a DNA sample. Prop. 69 thus radically changed the database from one comprising profiles of individuals *convicted* of violent felonies to one that includes profiles from *suspected* shoplifters.

These changes are leading to a huge increase in the number of people subject to DNA sampling. In 2008, about 53,000 people in California were convicted of offenses that would have qualified for inclusion in the database under the Pre-Prop. 69 version of the law.[11]But when all felony convictions are included, this number quadruples to more than 227,000.[12]

The 2009 expansion to include all persons arrested for felonies will drive these numbers ever further up if it is fully implemented.[13] In California, as discussed below, approximately 30 percent of individuals arrested for a felony are never convicted. For example, the California Department of Justice reports that of the more than 325,000 people arrested for felonies in California in 2008, some 97,500 were not convicted of any crime.[14] Another 13,000 to 19,000 were convicted only of a misdemeanor.[15] Thus, the change from a database containing people convicted of violent felonies to those merely arrested for *any* felony goes from 53,000 to 325,000.

This huge increase is not distributed equitably among all people. African Americans comprise 6.6 percent of California's population, but 22.6 percent of those arrested for felonies in the state.[16] Although, as discussed below, the possibility of race-based decision making at all levels of the criminal justice system makes it impossible to know whether changing from a database of people convicted of felonies to one including everybody arrested for felonies will result in an increase in the *proportion* of people of color in the database, it will clearly result in a significant increase in the absolute number of minorities included.

Once an individual's genetic information is entered into California's DNA database, it is difficult to get removed. An arrestee who is not convicted of a qualifying offense may petition for expungement of his or her DNA profile from the state's database.[17] However, the statute erects several procedural barriers. First, a person who is arrested but never charged with an offense can only file a petition after the legal time limit for charging the crime has expired.[18] This time limit, known as the statute of limitations, is at least three years for any felony; serious felonies have longer limitations periods or no statute of limitation at all.[19] An arrestee thus cannot even begin the process of asking to have her sample and profile removed for at least three years after the arrest.

Even after a petition is filed, the court cannot rule on it until at least 180 days have passed and the court is satisfied that neither the California Department of Justice nor the local prosecutor objects to expungement, effectively providing the government with an absolute veto.[20] People with prior felony convictions, no matter how old, are ineligible.[21] Even after surmounting all these obstacles, a person—even one who has been found innocent by a court—is not assured of success: the court has the discretion to grant or deny the request for expungement, and the decision is final, with no appeal allowed.[22]

This statutory scheme means that arrestees will have their DNA analyzed, databanked, and searched against the forensic database long before they can challenge it. Even if a person succeeds in obtaining a court order for the destruction of his sample, it is inconceivable that the government will truly expunge the profile not just from the current database but from the multiple backup copies that will have been made in the more than three years of delay.[23]

Moreover, for very practical reasons, few arrestees, no matter how innocent, will be likely ever to return to court to wade through the process of getting their samples and profiles removed. This is particularly true for those without the resources to hire a lawyer to assist them. Just as few Californians take advantage of existing procedures to have prior convictions expunged for employment purposes or records of wrongful arrests sealed—despite the obvious advantages of doing so—it seems likely that few will be both willing and able to have their DNA removed from the databank, particularly since the benefits of doing so may not be at all obvious. Because few people will have their DNA and profiles removed once they are included, the way we select people for initial inclusion becomes that much more important.

HOW RACIAL DISPARITIES ARE INTRODUCED INTO THE SYSTEM

Racial disparities fluctuate depending on the stage of the criminal proceeding, from the high-level initial decision to make certain acts criminal, to a police officer's decision to contact or arrest an individual, to the decisions made by prosecutors, judges, jurors, and defense lawyers. Thus, the stage at which DNA samples are taken will affect the racial disparities in the databank, albeit in unpredictable ways. The first and broadest stage at which racial disparities are introduced into the criminal justice system is at the legislative level, where crimes are defined and classified. The basic question is, of course, what conduct is considered criminal: why is a person who possesses drugs subject to criminal sanctions while a business that puts its workers or consumers at risk with dangerous or unsanitary plants subject only to civil sanctions? But even beyond that basic issue, our criminal justice system treats very similar conduct differently in ways that create racial disparities. The most notorious example of this is the disparity between the punishment for crack and powder forms of cocaine in the federal system, which for years punished people convicted of crack cocaine offenses (well over 80 percent of whom are African American) much more severely than powder cocaine offenders (72 percent of whom are white or Hispanic).[24]

Although this gross disparity was recently partially eliminated, the same pattern is repeated in other laws. California, for example, continues to punish possession for sale of crack more seriously than it does possession for sale of powder, although the disparity in the danger of the drugs is fairly minor.[25]

Much more significant is California's differential treatment of other classes of drugs. California has long divided its felony drug-possession offenses into two general classes: those punished under Health and Safety Code § 11350 (which include cocaine, cocaine base, and opiates ranging from heroin to codeine) and those punished under § 11377, most commonly methamphetamine, but also including amphetamine ecstasy (MDMA), LSD, and other less-common drugs. This distinction was first created when the state decided to criminalize the latter class of drugs in 1965.[26]

California's treatment of these two classes of drugs follows the same lines as the federal crack-powder disparity. Of the more than fifty-seven thousand adults arrested in 2008 for violating § 11377, 42 percent were classified as white and 8.5 percent as black, closely mirroring the state's general population.[27] In contrast, of the fifty-three thousand adults arrested under § 11350, 24 percent were white, 30 percent Hispanic, and 25 percent black. The percentage of black people arrested under § 11350, therefore, is four times the percentage of black people in the general population. Moreover, the legal consequences of being convicted for § 11350 are greater than those for § 11377. Simple possession of methamphetamine can be either a felony or a misdemeanor. In many parts of the state, possession of small quantities by people without extensive records will almost always be charged as a misdemeanor. Even when prosecutors charge it as a felony, the court may reduce it at any time up until conviction, or even after conviction if the defendant receives probation. Simple possession under § 11350, however, is always a felony. Although the government will on occasion agree to reduce such a charge to a misdemeanor, this is a rare act of grace. Even defendants who successfully complete probation will have a felony on their record for life, with all the consequences for employment, education, and licensing. And, most relevant to this discussion, they will have their DNA databanked with no possibility of having it removed.

The legislative establishment of "drug-free zones," often around schools, parks, or public-housing projects, can also have racially disparate effects.[28] These laws mean that people who live and commit drug crimes in dense urban areas, where few locations are *not* close to a school or park will be punished more harshly for the same conduct than are their suburban or rural counterparts. Because urban areas usually have higher proportions of people of color, these harsher punishments will reinforce racial disparities.[29]

Laws like these interact with seemingly race-neutral DNA collection laws to produce great disparities in the databank. A databank that includes all persons convicted of felonies will include every person, primarily people of color, convicted of possessing cocaine or heroin, no matter how small the amount. But the databanks will not contain samples from people who are primarily white, convicted of minor methamphetamine offenses that were prosecuted as misdemeanors. Conversely, a databank that includes only violent crimes or sex crimes, as many originally did, should result in fewer disparities than an all-felony database for the reasons just described. Excluding nonviolent crimes is reasonable since DNA evidence is almost never involved in nonviolent offenses. Thus, the move from databanks that include only serious crimes of violence to databanks that include all felonies will likely increase racial disparities.

Moreover, DNA databanks themselves create a feedback loop that further magnifies these disparities. Well over half of all serious crimes go completely unsolved, with the police never even identifying a suspect.[30] If DNA databanks work as they are intended, they will identify suspects for at least some, perhaps many, of these crimes. But a racially skewed databank will produce racially skewed results; because racial disparities in the criminal-justice system have led to the inclusion of a disproportionate number of profiles of African Americans in CODIS, the databank will return a disproportionate number of matches to African American suspects. In contrast, crimes committed by members of groups that are underrepresented in CODIS will escape detection, particularly as the police spend an increasing amount of their limited time and resources focusing on cases where they have found a DNA match.

The US Constitution does not prohibit this shift, regardless of the racial disparities it introduces. The courts have held that the Fourteenth Amendment's promise of equal protection of the law prohibits only intentional discrimination, which means that challenges to criminal laws that result in racially disparate impacts are extremely difficult. In the words of the US Supreme Court, discriminatory intent means "more than intent as volition or intent as awareness of consequences. It implies that the decision-maker, in this case a state legislature, selected or reaffirmed a particular course of action at least in part because of, not merely in spite of, its adverse effects upon an identifiable group."[31] If the body would have acted the same way even without the discriminatory intent, the law stands.

Thus, to invalidate the penalty disparities between different types of drugs, a court would have to determine that the legislatures enacted these laws in order to punish African Americans more harshly than whites rather than to deal with what was perceived as a dangerous epidemic related to the economics of the market and violence surrounding that market.[32] Even when the legislative record contains both overt and coded racial language, that is, references to an "invasion of Jamaican drug dealers" and "ghetto gangs" into the suburbs, courts are unwilling to find that the intent of the legislature as a whole was to discriminate.[33] Thus, after noting that the law was passed without much real consideration following the hysteria surrounding basketball star Len Bias's death (probably caused by powder rather than crack), one court noted that Congress's assumptions about the pharmacological differences between crack and powder have since been shown to be false. It stated that the "unjustifiably harsh crack penalties disproportionately impact on black defendants," and introduces "irrationality and possibly harmful mischief into the criminal justice system," but nonetheless concluded that "[o]nly Congress can correct the statutory problem."[34] Without a true smoking gun evincing overt racism, legal challenges to laws on the grounds that they are racially discriminatory are lost causes.[35] The US Court of Appeals has applied these same principles to reject an argument that the racial disparities in the federal DNA database made it unconstitutional.[36] No matter how disparate the impact of the database, without a showing that Congress enacted it to adversely affect African Americans, the challenge fails.

ALLOCATION OF POLICE RESOURCES

A second policy-level set of decisions also creates racial disparities: the allocation of law-enforcement resources. The clearest big-picture example of this is the so-called war on drugs, which is largely responsible for filling our prisons with men and women of color over the last thirty years.[37] A war on securities fraud or tax evasion would result in the arrest and prosecution of a very different demographic, namely, white males. But resources for combating these types of crimes have been cut, despite evidence that violations are common and devastating to our society, as evidenced by the current global impacts of finance fraud. On a smaller scale, police decisions to conduct "buy-bust" operations in specific neighborhoods, where undercover officers attempt to buy drugs from people on the street and then arrest anybody who sells them the drugs, mean that the police choose who will be targeted by what neighborhood is chosen for the operation.[38] Mirroring the emphasis on drug crime rather than white-collar crime, these operations usually occur in poor, urban neighborhoods with large minority populations. Because these types of mid-level resource-allocation decisions mostly affect the number of people legitimately arrested for crimes, they have the potential to create racial disparities that persist all the way through the criminal justice system. They will therefore affect the composition of DNA databanks regardless of what crimes are covered or at what stage of the process samples are taken.

RACIAL PROFILING BY INDIVIDUAL OFFICERS

Racial disparities also enter through racial profiling by individual officers. Studies have shown that some mixture of unconscious racism, conscious racism, and the middle-ground use of criminal profiles leads law enforcement to focus its attention and authority on people of color.[39] This can include everything from police officers disproportionately selecting people of color to approach, question, and ask consent to search, to discriminatory enforcement of traffic laws (e.g., the popular term "driving while black"), and detaining and arresting people of color without sufficient individualized suspicion.

The laxity of the legal requirements for this type of police action, coupled with a lack of supervision in the field, means that there are few checks against racial profiling. Police officers can lawfully approach and question a person without any reason to suspect that the person has done anything wrong, because *in theory* the person is free to ignore the officer and simply walk away. During one of these so-called consensual encounters, the police may lawfully ask the person to consent to a search.

Police officers do not need a lot of evidence to lawfully arrest a person: the standard is "probable cause," which means that an officer who reasonably believes that there is a "fair probability" that a person has committed a crime may arrest the suspect.[40] This standard allows the police to arrest people who may be innocent, or even those who are probably innocent. Despite the use of the term "probable cause," this does not mean that the person is probably guilty. The US Supreme Court has held, for example, that if the police find drugs in a car they can lawfully arrest all the occupants.[41]

In practical terms, of course, the police have even greater discretion to arrest people they suspect of being involved in criminal activity, even without probable cause. An officer may unlawfully take somebody into custody on a hunch. If the police subsequently uncover further evidence against the suspect, they will forward the case to the district attorney's office for prosecution. If the police do not uncover sufficient evidence to support prosecution and simply release the person after a day or two, they will rarely face any consequences. Unless the officer's actions were particularly egregious, few arrestees have the resources or the motivation even to file an internal-affairs complaint against the officer, much less bring a civil-rights lawsuit, particularly since very few lawyers will take such run-of-the-mill false arrest cases. Although the illegality of the arrest may—sometimes, if facts are clear—result in the suppression of a confession or other evidence that directly resulted from the wrongful arrest, it will not prevent prosecution based on other evidence. The police thus have a number of incentives, and few disincentives, to make an arrest even if they are not sure there is probable cause. As the current US Supreme Court continues to roll back the remedies available to criminal defendants who can show that the police violated their rights, these incentives will only expand.[42]

As with challenges to legislative actions, challenges to racial profiling under the US Constitution are extremely difficult because of the need to show discriminatory intent. This is magnified because the law gives police officers so much discretion as to who they will or will not approach, stop, question, or search. The US Supreme Court has held that the police may lawfully make pretextual stops, for example, singling out one driver among many who is speeding and stopping them because the officer has a hunch that they may be carrying drugs. This means that, although the police may not stop a person just because of race, in the unlikely event an officer is called upon to explain the stop, there is always something that can justify their actions: the driver or passenger's nervous glance at them, the driver reduced speed upon seeing the officer, a pedestrian wearing a heavy coat on a warm day, and other equally innocuous behavior. None of this behavior alone would justify the stop of a car, but such seemingly innocent actions are enough to justify the officer's decision to stop this particular car for driving a few miles per hour over the speed limit, while ignoring all the others that did the same, or to stop this individual for jaywalking while ignoring similar violations. Even if a court determines that an officer did make a stop because of nothing more than the driver's race, the only remedy is the possibility of a civil suit against the officer.[43] Unless serious harm was done, this is highly unlikely to occur so it is not a real deterrent to such police abuse of power.

These same considerations affect challenges to the way officers make so-called consensual encounters, ask drivers whom they have stopped for consent to search, detain them so that a drug-sniffing dog can come, or choose to arrest all of the occupants of a car where drugs are found rather than just the person closest to the drugs, the owner of the car, or the driver.[44] Unless the officer admits that s/he would not have acted the same way were the suspect of a different race, it is extremely difficult to prove that it was the cause.[45] In the absence of direct evidence of discriminatory intent, courts will occasionally accept statistical evidence of discriminatory enforcement. Again, though, the intent requirement has proved a nearly insurmountable obstacle. No matter how stark the statistics, the government can usually convince a court, that in an individual case, this particular stop was nonetheless not the product of intentional discrimination. Officers may simply testify that they made the stop for a valid reason, not because of race, and if the judge believes them that can end the matter. For example, the court in one reported case accepted an officer's testimony that he had stopped a car before seeing the race of the driver, and thus could not have been engaged in racial profiling, despite what the court called "disturbing" statistics that 34 percent of the officer's traffic stops had involved Hispanic drivers.[46]

More commonly, officers will testify that something caught their attention and led them to stop this particular person. Thus, statistics that the police in a train station approached and questioned only African American passengers over a ten-day period were deemed irrelevant when a federal agent testified that he had stopped the defendant because he had appeared nervous and the defendant could not produce any evidence that the agents had ignored white passengers who had similarly been "looking nervously over their shoulders."[47]

This means that the only cases where statistical proof of racial profiling is successfully presented are those where the government is not in a position to explain the stops. For example, in New Jersey, a study of traffic stops on freeways that showed racial profiling resulted in a series of court decisions leading to the dismissal of a number of cases.[48] A federal court in California allowed a civil suit filed by the American Civil Liberties Union of Northern California to go forward based on a statistical analysis of stops by the California Highway Patrol.[49] In both cases, the government was not in a position to try to explain the disparities.

The low level of proof required to make an arrest, combined with the difficulties of preventing arrests that are illegal for lack of proof or for discriminatory enforcement of laws, means that allowing DNA collection immediately after arrest will lead to large databases full of innocent people. Furthermore, given the ubiquity of racial profiling, people of color will largely populate the databases. The bottom line is that police end up with enormous discretion to determine who is in a database, with absolutely no review of many of their arrests.

GENETIC SURVEILLANCE

Arrestee sampling adds another incentive for the police to make questionable or outright illegal arrests. Whether or not the arrest leads directly to charges being filed, the arrestee's DNA profile will automatically be included in the database and run against all crime-scene evidence, now and in the future. Because of the barriers to having DNA samples removed, few arrestees will be able to have their samples and profiles expunged, thus allowing law enforcement the power to place people under lifetime genetic surveillance. To underscore, the consequence of the arrest of a plainly and indisputably innocent person will no longer be a short stint in jail, rather it will be a lifetime of genetic surveillance, turning the notion of a free and democratic society that protects civil liberties on its head.

Such surveillance constitutes a serious expansion of the consequences of an arrest and violates the Fourth Amendment to the US Constitution.[50] In recognition of the dangers to our liberty posed by the authority of a single officer to make an arrest, the US Supreme Court has held that the Fourth Amendment prohibits the police from holding a person in custody for more than forty-eight hours unless a judge determines that the facts set forth in sworn declarations show probable cause that the person has committed a crime.[51] Many states, including California, additionally require that a person arrested and held in custody be charged and brought to court to be arraigned within two working days of the arrest or else released.[52] These protections mean that arrestees do not spend more than a few days in jail without having both a judge and a prosecutor review the facts underlying the arrest.

This judicial and prosecutorial review results in more than fifty thousand arrestees in California—about 15 percent—being released from custody without ever even being charged with a crime each year.[53] A system that took DNA samples only after one or both of these levels of review would thus reduce the absolute number of people included in the system substantially. Moreover, the people thus excluded would be those who are most likely to be innocent of any crime. A police officer's incentive to make an arrest without probable cause in order to get a DNA sample would also be substantially reduced, because unless the officer took the additional step of writing a false report and presenting it to the prosecutor the person would likely be released without any sample being taken.

JUDICIAL AND PROSECUTORIAL POWER

The remaining 85 percent of arrestees are charged with crimes and begin what can be a long journey through the court system. The vast majority, close to 80 percent, end up pleading guilty to some crime as part of a negotiated plea. For those in custody on fairly minor charges who cannot raise bail, a quick plea bargain is often the only way they can avoid spending months in custody awaiting trial. Others seek a plea bargain to ensure a lighter sentence than they would receive if convicted at trial. A small percentage are able to take advantage of alternatives such as drug-treatment programs that divert them from the standard punishment track and often entitle them to a dismissal of all charges at the end of the treatment.

What all these defendants have in common is that their fates will be determined by the highly discretionary judgments of prosecutors, defense lawyers, probation officers, judges, and, for a very small percentage, jurors. A prosecutor who thinks the defendant is a menace may search the criminal code and add every possible charge and then take a similarly hard line in plea negotiations. Conversely, a prosecutor who is sympathetic to a particular defendant or just believes the police made a mistake, or that the defendant is not really the "criminal type" may offer to reduce a felony to a misdemeanor. In addition, a defense lawyer may fight harder for a client who can convince her that he is worth the extra trouble. On another level, probation officers recommend harsher or more lenient sentences based in part on the factors mandated by law but also on their overall estimate of a defendant's character and value to society. Judges may accept these recommendations or diverge from them because of any number of intangibles.

All of these human interactions require judgment and thus provide room for prejudgment as well. Not surprisingly, many studies have shown that all too often race becomes a factor.[54] But, once again, the courts' refusal to address anything other than intentional, individualized discrimination presents an insurmountable barrier to change. The most sophisticated attempt to address this problem occurred in the 1980s, in a case called *McCleskey v. Kemp*.[55] A Georgia court had convicted Warren McCleskey, who was African American, of killing a white police officer during a robbery and had sentenced him to death. Mr. McCleskey's lawyers presented the federal courts with a sophisticated study of two thousand murder cases in Georgia that, after taking into account 230 separate variables, showed racial disparities in the state's administration of capital punishment. Among the grim, uncontested numbers were comparisons of how differently a person charged with murdering an African American would be treated because of the defendant's race: the death penalty was imposed in 22 percent of these cases involving black defendants, as opposed to 1 percent where the defendant was white.

When the US Supreme Court decided the case, its response to these stark numbers was to raise the bar even higher; it stated that because so many decisions by so many actors based on so many different factors contribute to the final determination of whether a person will be sentenced to death, it is impossible to determine just where these disparities arise. Moreover, for the court to allow an investigation into the decision making of prosecutors, judges, and jurors would be an undue invasion into the discretion that our system accords these actors. Thus, ruled the Court, the study was "clearly insufficient to support an inference that any of the decision-makers in [the] case acted with discriminatory purpose."[56]

More telling, perhaps, is the Court's statement that accepting the "claim that racial bias has impermissibly tainted the capital sentencing decision, [it] could soon be faced with similar claims" in all sorts of criminal cases.[57] In other words, allowing a challenge to racially discriminatory imposition of the death penalty would require the Court to address the broader issue of racism in the criminal justice system, a step that the five-member majority was unwilling to take. As a dissenting justice wrote, the words of the majority opinion "suggest a fear of too much justice."[58]

CONCLUSION

The *McCleskey* case was decided in 1987, when the Supreme Court was still far less hesitant to step in to remedy what it saw as injustice, inequity, or unfairness in our criminal justice system than is the present Court. Twenty years later it still stands as a warning to those who look to the federal courts to protect our nation's civil rights and liberties. When states first started taking DNA from people convicted of serious crimes, there was some hope that the courts would bar the practice as unconstitutional. But in fact the federal appellate courts have unanimously, if over often-vigorous dissents, approved not only the collection of DNA from people convicted of violent felonies, but also laws mandating that people convicted of *any* felony, no matter how minor, provide samples.[59] Although the legal case against arrestee testing is much stronger, the few courts—mostly trial courts—that have addressed the issue have reached very different conclusions about its constitutionality.[60] Even if arrestee testing is invalidated and DNA is taken only after conviction of a crime, with all the procedural protections enjoyed by those in control of the criminal justice system, racial disparities will continue and accelerate as forensic DNA databases grow throughout the United States, unless the voters and legislatures put a stop to it.

NOTES

1. California Department of Justice, Division of California Justice Information Services, Bureau of Criminal Information and Analysis, Crime in California 2008 Data

Tables, table 37, http://ag.ca.gov/cjsc/publications/candd/cd08/preface.pdf (hereinafter "Crime in California"). The year 2008 is the most recent for which complete data are currently available.

2. Between 2002 and 2009, the number of states that require persons convicted of felonies to provide a DNA sample has gone from twenty-two to fifty. *Compare* State Laws on DNA Data Banks Qualifying Offenses, Others Who Must Provide Sample (2009) *with* Fighting Crime with DNA (2002), both available from the National Conference of State Legislatures at www.ncsl.org/programs/cj/dna.htm.
3. *See* Cal. Penal Code § 295(e), 28 C.F.R. § 28.12(f)(1); 73 Federal Register 74935 (December 10, 2008) ("[T]he states that collect DNA samples from arrestees typically do so by swabbing the inside of the person's mouth ('buccal swab'), and many states use the same method to collect DNA samples from convicts.")
4. A description of the CODIS system can be found on the federal government's DNA Initiative website, www.dna.gov/dna-databases/codis.
5. DNA Identification Act of 1994, Pub.L. 103–322, Title XXI, § 210304, Sept. 13, 1994, 108 Stat. 2069, codified at 42 USCA § 14132. One consequence of this federal program was to make it easier and cheaper for states to develop their own programs for collecting and databanking DNA—the FBI provides the CODIS software for free—as well as to make individual state DNA collection and databanks more useful, because they would be linked to the national databank. The FBI also maintains the National DNA Index System (NDIS), which comprises all the DNA profiles in the CODIS system and automatically searches them every week to match crime-scene evidence to profiles taken from known individuals or to evidence from other crime scenes (www.dna.gov/dna-databases/codis).
6. Law enforcement have also begun to use the databanks for so called "familial searches," where a forensic sample is similar to a known sample, on the theory that it is likely that the culprit is related by blood to the person who provided the known sample. *See* Henry T. Greely, Daniel P. Riordan, Nanibaa' A. Garrison, Joanna L. Mountain, *Family Ties: The Use of DNA Offender Databases to Catch Offenders' Kin, Journal of Law, Medicine & Ethics* 34 (Summer 2006): 248–62. Recently both the California and federal Departments of Justice have authorized the use of partial matches to conduct so-called familial searching. California Department of Justice, Division of Law Enforcement Information Bulletin 2008-BFS-01, DNA Partial Match (Crime Scene DNA Profile to Offender) Policy (April 24, 2008), http://ag.ca.gov/cms_attachments/press/pdfs/n1548_08-bfs-01.pdf; CODIS Bulletin BT072007, Interim Plan for Release of Information in the Event of a "Partial Match" at NDIS (July 20, 2006), www.bioforensics.com/conference08/Familial_Searches/CODIS_Bulletin.pdf; *see* Maura Dolan and Jason Felch, "Tracing a Crime Suspect Through a Relative," Los Angeles Times, November 25, 2008.
7. Former Cal. Penal Code § 290.2, enacted by Stats.1989, ch. 1304, § 1.5, pp. 5176–5178. For a discussion of the enactment, amendment, and eventual replacement of this provision see People v. King, 82 Cal.App.4th 1363, 1369–70 (2000).
8. Former Cal. Penal Code § 290.2; *see People v. King, supra*, 82 Cal.App.4th at 1369–70.

9. The list of qualifying offenses was expanded from murder and specified sex offenses to include voluntary manslaughter, felony domestic violence, aggravated assault, kidnapping, mayhem, torture, residential burglary, robbery, arson, and carjacking. See Cal. Penal Code § 296 historical note; Stats.1996, ch. 917, § 2; *People v. Brewer*, 87 Cal.App.4th 1298. *See also* CA LEGIS 696 § 2 (1998) (enacting PC 295(b)(3),(c)).
10. Cal. Health and Safety Code §§ 11350, 11377 (felony drug possession); Cal. Penal Code §§ 459 (felony to enter a store with the intent to shoplift), 666 (second-time shoplifting is a felony), 476a (felony to present check with insufficient funds).
11. See Crime in California, supra, tables 3, 8, 42. This figure is estimated from the number of convictions for violent crimes added to the number of residential burglary convictions, calculated as 22.5 percent of the number of convictions for property crimes based on data in tables 3 and 8 (36 percent of property crimes are burglaries, of which 62.5 percent are residential burglaries). Because these data do not distinguish between felony and misdemeanor convictions, they overstate the number of people who are convicted of felonies. Some of this overcount can be corrected by subtracting the number of people sentenced only to jail, which indicates that the person was convicted of a misdemeanor. *See* Cal. Penal Code § 17.

    California's total population is 36.96 million people, according to the most recent estimate by the US Census. See http://quickfacts.census.gov/qfd/states/06000.html.
12. *See* Crime in California, table 37. This ratio comports with national statistics; nationally, less than 20 percent are violent crimes or residential burglaries. United States Department of Justice, Bureau of Justice Statistics, State Court Sentencing of Convicted Felons, 2004—Statistical Tables, table 1.1, http://bjs.ojp.usdoj.gov/index.cfm?ty=pbdetail&iid=1533.
13. There are serious questions about whether compulsory seizure of DNA from people who have merely been arrested for crimes violates the Fourth Amendment to the US Constitution, which prohibits unreasonable searches and seizures, and several challenges to such laws are pending in the courts. The United States Court of Appeal for the Ninth Circuit—which includes California—has ruled that the forcible extraction of DNA from an arrestee, without a warrant or an authorizing statute, violated the Fourth Amendment. *Friedman v. Boucher*, 580 F.3d 847 (9th Cir. 2009). The Minnesota Court of Appeals, as well as two federal trial courts, have invalidated arrestee-collection laws for this same reason. *In re Welfare of C.T.L.*, 722 N.W.2d 484 (Minn. App. 2006); *United States v. Mitchell*, 681 F.Supp.2d 597, (W.D.Pa. 2009) (appeal pending); *United States v. Purdy*, 2005 WL 3465721 (D.Neb. 2005). In contrast, The Virginia Supreme Court has upheld that state's arrestee-testing law, and a federal trial court refused to issue a preliminary injunction to halt California's new law. *Anderson v. Commonwealth*, 650 S.E.2d 702 (Va. 2007); *Haskell v. Brown*, 687 F.Supp.2d 1187, (N.D. Cal. 2009) (appeal pending). The United States Court of Appeals for the Ninth Circuit recently held that the government may take DNA from a person accused of a crime after a judge has determined that there is probable cause that the person is guilty. *United States v. Pool*, 621 F.3d 1213 (9th Cir. 2010)

(petition for review *en banc* pending) Appeals or petitions for rehearing are pending in all three of the recent federal cases.

14. Crime in California, *supra*, table 37.
15. *Id.* table 42.
16. Crime in California, *supra*, table 31; US Census quick facts, *supra*.
17. Cal. Penal Code § 299.
18. *Id.* § 299(b)(1) (person arrested but not charged may only file petition if "no accusatory pleading has been filed within the applicable period allowed by law.").
19. *Id.* §§ 799–801.
20. *Id.* § 299(c)(2)(D).
21. *See id.* § 299(b).
22. *Id.* § 299(c)(1); *see Id.* § 299(b)(3) (procedure applies to those previously found factually innocent).
23. In 2006, the FBI reported to the US Department of Justice Inspector General that it backs up the national database "routinely" onto tapes. Audit Report 06–32, Combined DNA Index System Operational and Laboratory Vulnerabilities, Findings § 1 (May 2006), www.usdoj.gov/oig/reports/FBI/a0632/findings.htm#ID.
24. The law and figures are summarized in "Crack vs. Powder Cocaine: a Gulf in Penalties," *U.S. News and World Report*, October 1, 2007.
25. Possession for sale of crack cocaine carries a sentence of three to five years, of powder cocaine two to four years. Cal. Health and Safety Code §§ 11351, 11351.5.
26. *See* former Cal. Health & Safety Code § 11910, added by Stats.1965, c. 2030, p. 4603, § 1; see generally *Baker v. Superior Court*, 24 Cal.App.3d 124, 126 (1972). Heroin and cocaine had been illegal since 1939. See former Health & Safety Code § 11500, enacted by Stats.1939, c. 60, p. 758.
27. Crime in California, *supra*, table 31. In 2009, California's population was 6.6 percent African American, 41.7 percent white. US Census quick facts, *supra*.
28. *See, e.g.*, Cal. Health & Safety Code § 11353.6 (adding addition sentence of 3–5 years for specified drug offense committed within 1000 feet of a school); *id.* § 11380.7 (additional sentence of 1 year for specified drug crimes committed within 1000 feet of a homeless shelter or drug program).
29. This issue is discussed at length in a 2006 report by the Justice Policy Institute. *See* Judith Greene, Kevin Pranis and Jason Zeidenberg, "Disparity by Design: How Drug-Free Zone Laws Impact Racial Disparity—and Fail to Protect Youth" (2006), www.justicepolicy.org/images/upload/06-03_REP_DisparitybyDesign_DP-JJ-RD.pdf.
30. According to the FBI, "Nationwide in 2007, law enforcement cleared 44.5 percent of violent crimes and 16.5 percent of property crimes" by identifying or arresting a suspect. United States Department of Justice—Federal Bureau of Investigation, "Crime in the United States, 2007," www.fbi.gov/ucr/cius2007/offenses/clearances/index.html.
31. *Personnel Adm'r of Massachusetts v. Feeney*, 442 U.S. 256, 279 (1979).

32. *United States v. Williams*, 962 F.2d 1218C.A.6 (Mich.),1992. at 1227. The court wrote that, "Congress was clearly concerned that the special attributes of crack—its small size and cheap price per dose—could create other societal problems that required remedying. Senators noted that because crack is sold in small doses (called 'rocks') it is easier to transport and use, thereby increasing the difficulty of suppressing addiction. The cheap price of each 'rock' also permits children to afford cocaine for the first time, thereby exposing another segment of American society to drug addiction."
33. *United States v. Walls*, 841 F.Supp. 24 (D.D.C. 1994).
34. *United States v. Smith*, 359 F.Supp.2d 771 (E.D.Wis. 2005).
35. *United States v. Clary*, 34 F.3d 709, C.A.8 (Mo.),1994.
36. *Wilson v. Collins*, 517 F.3d 421 (6th Cir. 2008).
37. *See* Ryan S. King, "Disparity by Geography: The War on Drugs in America's Cities" (2008), www.sentencingproject.org/Admin%5CDocuments%5Cpublications%5Cdp_drugarrestreport.pdf.
38. *Id.* at 21.
39. *See, e.g.*, Katherine Beckett, "Race and Drug Law Enforcement in Seattle," www.prisonpolicy.org/scans/Beckett-20040503.pdf; Ian Aryes, "Racial Profiling & The LAPD: A Study of Racially Disparate Outcomes in the Los Angeles Police Department," www.aclu-sc.org/documents/view/47.
40. *United States v. Buckner*, 179 F.3d 834, 838 (9th Cir. 1999).
41. *Maryland v. Pringle*, 540 U.S. 366 (2003).
42. *See, e.g., Herring v. United States*, 129 S.Ct. 695 (2009) (refusing to exclude unlawfully obtained evidence because police acted negligently, not intentionally); *Hudson v. Michigan*, 547 U.S. 586 (2006) (refusing to exclude evidence that police obtained by unlawfully breaking down door of house). A majority of the court has suggested that the increase in professionalism among police departments may mean that the exclusionary rule may no longer be necessary. *See id.* at 598–99.
43. *United States v. Nichols*, 512 F.3d 789 (6th Cir. 2008).
44. *Farm Labor Organizing Committee v. Ohio State Highway Patrol*, 308 F.3d 523 (6th Cir. 2002).
45. Although officers sometimes do admit that they treat people of color differently from whites or exhibit such overtly racist behavior that they cannot later hide their intent, such direct evidence of discriminatory intent by law enforcement "seldom exists." *United States v. Avery*, 137 F.3d 343, 355 (6th Cir. 1997). For examples of cases where plaintiffs were able to present such direct evidence, see Farm Labor Organizing Committee v. Ohio State Highway Patrol, 308 F.3d 523 (6th Cir. 2002) in which officers "all testified that, in their experience, they would refer Hispanic motorists to the Border Patrol when, in precisely the same circumstances, they would not refer someone who was white (i.e., not of Hispanic appearance") and *Carrasca v. Pomeroy*, 313 F.3d 828 (3d Cir. 2002) where the plaintiffs argued that the officer's use of a racial slur, combined with the evidence he had singled the plaintiffs out for arrest, showed an intent to discriminate. In both cases the courts held that the plaintiffs had

presented sufficient evidence to let a jury determine whether the police had been motivated by bias.

46. *United States v. Alcaraz-Arellano*, 441 F.3d 1252 (10th Cir. 2006).
47. *United States v. Barlow*, 310 F.3d 1007 (7th Cir. 2002). For a similar analysis by Massachusetts's highest court *see Commonwealth v. Lora*, 451 Mass. 425 (2008).
48. *State v. Soto*, 734 A.2d 350 (N.J. Super 1996).
49. Rodriguez v. California Highway Patrol, 89 F.Supp.2d 1131 (N.D.Cal. 2000).
50. *See In re Welfare of C.T.L.*, 722 N.W.2d 484 (Minn. App. 2006) *and United States v. Purdy*, 2005 WL 3465721 (D.Neb. 2005) (both holding that arrestee testing violates the Fourth Amendment).
51. *County of Riverside v. McLaughlin*, 500 U.S. 44 (1991).
52. Cal. Penal Code § 825.
53. Crime in California, *supra*, table 38.
54. Marc Mauer, *Racial Impact Statements as a, Means of Reducing Unwarranted Sentencing Disparities*, 5 Ohio State Journal of Criminal Law 19, 27 (2007).
55. *McCleskey v. Kemp*, 481 U.S. 279 (1987).
56. *Id.* at 297.
57. *Id.* at 314.
58. *Id.* at 339 (Brennan, J., dissenting).
59. *See, e.g., United States v. Kriesel*, 508 F.3d 941, 946–47 (9th Cir. 2007) (collecting cases); *United States v. Weikert*, 504 F.3d 1, 10 (1st Cir. 2007).
60. See note 13, *supra*.

4

PREJUDICE, STIGMA, AND DNA DATABASES

Helen Wallace

The collection, use, and storage of DNA for forensic purposes have increased rapidly since 1995, when the world's first DNA database was set up in Britain. The use of DNA in criminal investigations can undoubtedly be highly beneficial, providing evidence that can help to convict the guilty and exonerate the innocent. Storing DNA evidence from crime scenes and the computerized DNA profiles obtained from it can also be extremely valuable if past crimes need to be reinvestigated. However, the retention of individuals' DNA profiles and other information on computer databases, combined in some countries with the storage of linked biological samples, raises many privacy and individual rights issues. These include how the information might be misused by governments or others, and the prospect that false DNA matches may lead to intrusive investigations of innocent people by law enforcement agencies.

This chapter discusses the evidence and reasons that innocent people whose DNA profiles are contained in DNA databases may be vulnerable to stigmatization or prejudicial treatment. It draws on experiences from the development and operation of the National DNA Database in Britain, which contains the largest proportion of DNA samples of the population of any DNA database in the world.

WHAT IS SPECIAL ABOUT DNA?

DNA and fingerprints differ from other means of surveillance, such as photographs and iris scans, because they do not require equipment to be installed in particular places (such as a border control) in order to trace or record where an individual has been. Both DNA and fingerprints may be left wherever a person goes. The retention of DNA profiles and fingerprints from an individual on a database therefore allows a form of biological tagging or "biosurveillance," which can be used to establish whether they have been present at a particular location.[1] This purpose goes beyond mere "identification," to mapping an individual's movements, including (but not limited to) using biological evidence to establish their likely presence at a crime scene.

Unlike fingerprints, DNA can also be used to investigate biological relationships between individuals (including paternity and nonpaternity), and thus trace other individuals who may be related to a person whose DNA profile has been obtained from a crime scene or elsewhere. Biological relationships can be statistically inferred from computerized DNA profiles by searching for partial matches between profiles (an indication of relatedness), a process known as "familial searching."

The computerized DNA profiles held in DNA databases are a string of numbers based on specific areas of each individual's DNA, known as short tandem repeats (STRs). However, some countries such as Austria, France, and the United Kingdom also retain the biological samples collected from individuals, which are linked to their record on the computer database by a reference number. A person's DNA sample contains additional private information about their health and other physical characteristics. Some of this information (such as carrier status for a genetic disorder) may be highly sensitive or unknown to the individual. Health information can currently only be gleaned from biological samples by undertaking additional genetic analysis of stored samples, not from the DNA profiles themselves.

Although DNA has some special characteristics, the extent to which retained DNA profiles may lead to stigmatization or prejudicial treatment also depends on the extent to which other information (including names, addresses, ethnic appearance, and suspected crime) is retained alongside the DNA profile, or in linked computer databases, and how this information may be accessed and used.

THE NATIONAL DNA DATABASE IN BRITAIN

The National DNA Database (NDNAD; hereafter "Database") in Britain was the first to be established and contains a much larger proportion of the population than any other country in the world. An estimated 576,250 individuals had records added to the Database in 2006–07: one person every minute.[2] About 4.9 million people—nearly 8 percent of the population—had their DNA profiles retained on the Database by the end of December 2009.[3] Approximately 1 million of these individuals have never been convicted or cautioned for any crime. Cautions are issued at the police station for minor offenses and do not require a person to be charged or convicted by a court. Many countries are considering establishing or expanding their databases in line with the changes made in Britain, and examination of the NDNAD therefore provides an opportunity to consider the potential for stigmatization or prejudicial treatment of individuals with records on the Database.

LEGAL FRAMEWORK

There is no specific piece of legislation governing the operation of the National DNA Database. Instead, a series of laws have established the circumstances under which the police may take and retain DNA samples and data.[4] The Database was set up in 1995, but in recent years it has expanded rapidly due to two changes in the law.

In 2001, legislation was introduced as part of the Criminal Justice and Police Act to allow DNA profiles to be kept on the Database even when a person was acquitted of a crime. This change in legislation applied in England, Wales, and Northern Ireland. It was not fully implemented in Northern Ireland, although an agreement was adopted allowing the export of individuals' DNA profiles from Northern Ireland to the NDNAD.[5] An estimated fifty thousand profiles from acquitted persons may have been kept illegally on the Database before the law was changed;[6] the 2001 Act appears to have been intended to bring the law in line with this unlawful practice. There was no separate vote or debate on the part of the bill relating to the retention of DNA from innocent people and the UK Government was criticized by Members of Parliament (MPs) for introducing the 2001 legislation without allowing sufficient time for debate.[7]

In April 2003, the law was changed again to allow DNA to be taken as soon as a person is arrested, rather than waiting for them to be charged with an offense: this legislation came into affect in England and Wales in April 2004. The section of the Criminal Justice Act 2003, which allows DNA to be taken on arrest, rather than on charge, was introduced via a late amendment submitted by the Secretary of State during the first week of the Iraq war. This section of the bill required a separate vote in the House of Commons. No Northern Ireland MP from any party voted in favor of it; however, the provisions were later applied to Northern Ireland via the Criminal Justice (Northern Ireland) Order 2004, using special powers granted to the Secretary of State while the Northern Ireland Assembly was suspended.[8]

In England, Wales, and Northern Ireland, the police now take DNA samples routinely without consent from anyone aged ten or over (the age of criminal responsibility) who is arrested in connection with any recordable offense and taken to a police station. The police may use "reasonable force" (which usually involves pulling out a few hairs from a person's head) if the arrested person refuses to allow a cheek swab to be taken. Recordable offenses include begging, being drunk and disorderly, taking part in an illegal demonstration, and minor acts of criminal damage caused by children kicking footballs or throwing snowballs. In Scotland, DNA is taken on arrest for "imprisonable" offenses. This is a narrower category of offenses than in the rest of the United Kingdom, but still includes minor offenses such as "Breach of the Peace."

Uses of the NDNAD may include any purpose "related to the prevention or detection of crime." Uses now include familial searching (using partial DNA matches to try to identify the relatives of a suspect), searching by name, and undertaking various types of genetic research (including controversial attempts to predict ethnic appearance from DNA).[9]

INDEFINITE RETENTION OF DNA PROFILES FROM UNCONVICTED PERSONS

From 2001 to 2010, the law in England, Wales, and Northern Ireland allowed the police to retain all DNA profiles permanently, whether or not a person was charged or convicted. This was out of step with practice in other European countries and with the principles adopted by bodies such as the Council of Europe,[10] which require time limits on the retention of DNA profiles for all but the most serious offenders.

The indefinite retention of DNA profiles from innocent persons on the NDNAD was controversial and has been criticized by many, including the influential Nuffield Council on Bioethics.[11] This practice was subject to challenge in the European Court of Human Rights and, in December 2008, the Court ruled unanimously that the UK Government was in breach of article 8 of the European Convention on Human Rights (the right to privacy).[12] In April 2010, new legislation was adopted in England, Wales, and Northern Ireland allowing the retention of unconvicted persons' DNA profiles for six years following arrest (or for three years in the case of juveniles) rather than indefinitely.[13] This law was adopted by parliament but never brought into force due to a change in government. The new Government has promised to go further in protecting civil liberties and bring the law in the rest of the United Kingdom into line with Scotland. However, its new Freedom Bill had not been published at the time of writing. In May 2006, the Scottish Parliament voted against indefinite retention of DNA profiles and samples from persons acquitted or not proceeded against.[14] Instead, police powers were expanded to allow temporary retention (for up to three years, extendable by periods of two years with judicial oversight) from a much smaller number of people who had been charged but acquitted of a serious violent or sexual offense.[15]

RETENTION OF ADDITIONAL INFORMATION AND BIOLOGICAL SAMPLES

In England and Wales, records on the NDNAD include a person's name and ethnic appearance, as well as DNA profile. Ethnic appearance is categorized according to a government coding system, based on appearance to a police officer rather than an individual's own description of their ethnicity. The categories are Afro-Caribbean, Arab, Asian, Dark-skinned European, Oriental, and White-skinned European.

Operation of the National DNA Database transferred from the Forensic Science Service (FSS) to a new government agency, the National Policing Improvement Agency (NPIA), in 2007. A new forensic regulator and ethics board were also established, however neither has any statutory powers. The FSS was made into a government-owned company in 2005, and further privatization is under consideration. The FSS now competes with other commercial companies to analyze and store DNA samples for the police.

Part of each sample collected by the police is kept permanently by whichever company has analyzed it, for an annual fee. The sample is linked to the individual's record on the DNA Database using a unique barcode reference number. People who have been arrested also have an arrest summons number (ASN) included in their record on the NDNAD, which provides a link to other information on the Police National Computer (PNC). When the NDNAD was established in 1995, records were supposed to be removed at the same time as an individual's criminal record.[16] However, the change in legislation allowing DNA records to be retained has subsequently been used to justify a change in policy that requires keeping all PNC records permanently.[17] The retention of permanent records of arrest is unprecedented in British history. Retention of the PNC records enables the police (who do not have direct access to the NDNAD) to establish whether or not a DNA sample has already been taken from an arrested person. However, PNC records may also be accessed by a much wider range of individuals and agencies than the DNA Database and used for other purposes, such as preemployment checks. The Home Office is in the process of establishing a Police National Database (PND), which will provide a single access point for searching across all of the information currently held on the PNC.

WHO IS ON THE NATIONAL DNA DATABASE AND WHY?

Before considering how being on the database may lead to stigmatization or prejudicial treatment, it is important to consider who has their

TABLE 4.1 ESTIMATED NUMBERS OF INDIVIDUALS ON THE NDNAD AT END JUNE 2006

Unconvicted persons with a PNC record	605,069
Persons who have received noncustodial sentences or cautions, recorded on the PNC	1,681,284
Persons who have had a custodial sentence, recorded on the PNC	636,271
Estimated total no. individuals on NDNAD with a PNC record	2,922,624
Estimated total no. of individuals on NDNAD with no PNC record (includes 18,056 volunteers)	534,376*
Estimated number of individuals with profiles on the NDNAD	3,457,000*
Estimated number of replicate profiles	427,270*
Total number of individuals' DNA profiles on the NDNAD	3,884,270

*Assuming the 11 percent replication rate then in use
Source: Author

DNA profile recorded on the Database and why. The number of individuals with their data held is not known exactly, because a number of records appear to be duplicates of the same DNA profile, associated with different names. The latest estimates of the breakdown of individuals on the Database are given in table 4.1, which is based on figures given in response to a Parliamentary Question (PQ).[18] (PQs are asked by members of parliament to obtain official information and answered by government ministers.)

More recent figures on the number of people with noncustodial sentences or cautions have not been released, but in April 2009 an estimated 986,185 persons had DNA records on the NDNAD but no record of conviction or caution on the PNC.[19] Unconvicted persons arrested in 2006 or later will have a PNC record, but those arrested before this date may not. Most of these million people will have been acquitted or not charged, although some will be awaiting trial. The majority will have had their DNA taken on arrest (or, before 2004, on charge) without their consent; only the 18,000 volunteers are required to give consent. A further 1.68

million people with DNA profiles retained on the Database have received noncustodial sentences or cautions based on the 2006 data.

Most individuals do not have their DNA profile entered on the DNA Database to investigate a specific crime, since DNA evidence is relevant to less than 1 percent of crimes. Because DNA is taken routinely on arrest for any recordable offense, the purpose is usually to include the individual's DNA profile on the Database in order to perform a "speculative search" and look for matches with DNA profiles from all past unsolved crimes. Whatever the outcome of the speculative search, and regardless of whether a person is charged, cautioned, or convicted, DNA profiles are then retained and subjected to all future speculative searches of the Database. Currently, the policy to retain all profiles until the individual reaches age one hundred. A match between the individual's DNA profile and a DNA profile obtained from the scene of a *future* crime may then lead to the individual being identified as a possible suspect for that crime. Match reports, which identify one or more suspects for a crime, are sent to the police force that submitted the original crime scene DNA sample. The "added value" of putting individuals on a database is only to introduce new suspects into an investigation, not to exonerate innocent individuals who are already suspects for a crime (from whom DNA can always be taken in relation to that crime, without entering their details on a database).

Contrary to popular perception, the figures on DNA matches are dominated by volume crimes, such as burglaries and theft, not by more serious crimes such as rape and murder, for which the Database is much less effective. The proportion of DNA detections is much higher for these volume crimes, for which the value of using the Database is greater. The main reasons for this are that detection rates have historically been low for crimes such as burglary and that the identity of the suspect is often unknown, so a "cold hit" from the database can provide an important lead for an investigation.[20] The Database (as opposed to the use of DNA evidence in court) is much less effective for violent crimes because: (1) Most murderers, rapists, and perpetrators of violent crimes such as assault are known to their victims. This means that, although DNA evidence can sometimes be an important part of the evidence in court, the Database is not usually necessary to identify a list of suspects. (2) The most useful DNA in murder investigations is often the victim's DNA, not the perpetrator's. For example, scientists might examine bloodstains on someone's clothing to see if they came from the victim. Whether the suspect's DNA is on the Database or not is not relevant to this comparison. (3) In some types of cases, such as rape, there is often no disagreement about identity (i.e., the man involved) but a disagreement about whether a crime has taken place (whether the woman has given her consent). DNA can help solve disputes about identity, but not about consent. Although Britain has by far the largest DNA database in Europe, it has the lowest conviction rate for rape.

RACIAL BIAS

Approximately 27 percent of the entire black population, 42 percent of the male black population, 77 percent of young black men, and 9 percent of all Asians have records on the National DNA Database, compared with just 6 percent of the white population.[21] These figures are approximate because they are calculated by comparing the proportion of the population recorded as "Afro-Caribbean" on the Database (based on appearance to a police officer) with the proportion identifying themselves as belonging to the relevant ethnic group in the national census.[22] According to the Liberal Democrat MP Sarah Teather, an estimated 55 percent of those from black and minority ethnic communities who have their details on the database have never been charged or convicted of any offense.

The figures undoubtedly reflect both social exclusion and discrimination in the criminal justice system. A 2007 parliamentary report concluded that statistics show that young black people are overrepresented at all stages of the criminal justice system.[23] Black people constitute 2.7 percent of the population aged ten to seventeen, but represent 8.5 percent of those of that age group arrested in England and Wales. As a group, they are more likely to be stopped and searched by the police, less likely to be given unconditional bail, and more likely to be remanded in custody than white young offenders. Young black people are overrepresented as suspects for certain crimes such as robbery, drug offenses, and, in some areas, firearm offenses. Young black people and those of "mixed" ethnicity are also likely to receive more punitive sentences than young white people and more likely to be victims of violent crimes.

The report concluded that social exclusion, educational underachievement, and school exclusion interact to form a web of disadvantage, bringing young black people disproportionately into contact with crime and the criminal justice system as both victims and offenders. The report also argues that the relationship between black communities and the police in Britain leads to greater involvement in the criminal justice system, in some instances due to discrimination, and in other cases because suspicion or mistrust of criminal justice agencies leads young people to take the law into their own hands to protect themselves or exact redress.

A 2003 Home Office crime study found that for every age group, there was either no statistical difference among white, mixed race, and black groups or, more commonly, that white people were more likely to have offended than either mixed race or black people.[24] The report was based on self-reporting of offending from twelve thousand individuals up to the age of sixty-five. Over 42 percent of white people reported that they had committed an offense in their lifetime, against 28 percent of black people, and 21 percent of white people reported committing a serious offense in their lifetime, against 14 percent of black people.

CHILDREN AND YOUNG PEOPLE

More than one million of the four million profiles on the National DNA Database in 2007 were taken from children and young people who were under eighteen years of age at the time of their arrest.[25] DNA may also be taken from children under ten if their parents give consent. In England and Wales, the Crime and Disorder Act 1998 removed the presumption that a child aged ten to fourteen was *doli incapax* (incapable of wrong), which had required the prosecution to prove that the child knew the act to be seriously wrong, rather than merely naughty. This is one of the lowest ages of criminal responsibility in Western Europe. For the purposes of taking DNA, children who are aged ten or above are therefore treated in the same way as adults. Freedom of Information requests by Action on Rights for Children (ARCH) have confirmed that police forces in England and Wales have no special policies regarding the collection or retention of DNA from children.

Along with the reduction in the age of criminal responsibility, police targets for arrests and convictions have led to significant increases in the numbers of arrests of children and young people. The police themselves have raised many concerns about this "target culture." The Police Federation has described some arrests, such as that of a child for throwing a slice of cucumber, as "ludicrous."[26] Chief Inspector Sir Ronnie Flanagan, who recently published a review of policing, raised concerns with the Home Affairs Committee of MPs that children were being arrested for fighting in school playgrounds.[27] A child who has done nothing wrong can also be arrested in these situations, because one of the children will often make a counteraccusation, which may or may not be true. Both children will then have their DNA taken and entered on the Database. Police Constable Stuart Davidson told the BBC, "We get exactly the same points for cautioning a girl for pulling another girl's hair as we would for domestic burglary. In terms of statistics they're exactly the same."[28] Police Commander Brian Paddick, then running for Mayor of London as a candidate of the Liberal Democrat party, also expressed concerns about the points system to the Home Affairs Committee: "Clearly that is a nonsense, and clearly it is distorting what the police are concentrating on."[29]

The former chair of the Youth Justice Board, Professor Rod Morgan, summarized the situation:

> To meet crime targets, the police are picking low-hanging fruit—the lowest of which comprises juvenile group behavior in schools, residential homes and public spaces, offences that could be dealt with informally, more effectively, speedily and cheaply, and in former times were. There has been a 26% increase in the number of children and young persons criminalized in the past three years. This at a time when the British Crime Survey and police statistics indicate that most crimes, including those committed by juveniles, have been falling.[30]

According to Action on Rights for Children, government figures show that the arrests of those ten to seventeen years old are rising disproportionately: between 2002 and 2006 they rose by 16.4 percent to 348,500, while adult arrests rose by 6.6 percent during the same period.[31] As the number of arrests has risen, so too have arrests that do not lead to any disposal. In 2003–04, 25,317 arrests of children and young people received no further action. By 2005–06, this had risen by 84 percent to 46,640.[32] Exact figures are not available for the number of those under eighteen years old with records on the National DNA Database who have not been convicted of any offense, but GeneWatch UK and Action on Rights for Children have estimated that about one hundred thousand children and young people fall into this category.[33] This figure does not include children given reprimands and final warnings by the police. Reprimands and final warnings account for around 44 percent of disposals, and some parents may underestimate the seriousness of these sanctions and fail to ensure that their children get legal advice at the police station; they may even exert pressure on children to make admissions in order to "get it over with." A reprimand or final warning is not a finding of guilt in law, and they can be administered without the consent of the child or their parent.

IMPLICATIONS FOR INDIVIDUALS WITH DNA PROFILES ON THE DATABASE

The NDNAD is a useful tool in criminal investigations, but the retention of DNA from everyone who has been arrested for a recordable offense raises important concerns about privacy and rights, including:

the potential threat to "genetic privacy" if information is revealed about health or family relationships, not just identity;
the creation of a permanent "list of suspects" that could be misused by governments or others;
the potential for unauthorized access, abuses, and misuses and mistakes; and
the exacerbation of discrimination in the criminal justice system.

These vulnerabilities give rise to a number of places where prejudicial treatment can occur toward individuals who have had their DNA profiles included in the database. Prejudicial treatment and stigma can arise due to:

false matches with crime scene DNA;
the potential for harassment or stigmatization of innocent people or minor offenders;

the possible use of the database for the tracking or surveillance of individuals or their relatives;
the risk that people with records on the database will be refused visas or a job as the result of a record of arrest; and
the loss of an individual's right to refuse to take part in controversial genetic research.

Each of these is discussed in more detail below.

FALSE MATCHES

The use of DNA in criminal investigations can undoubtedly be highly beneficial, providing evidence that can help to convict the guilty and exonerate the innocent. However, DNA evidence is not foolproof and individuals who have their DNA profiles retained on a Database may face being wrongly implicated in a crime because of a false match between their profile and a DNA profile collected from a crime scene, or perhaps because they were present at the crime scene earlier in the day, but were not the perpetrator.

Many DNA matches are not with the perpetrator of a crime. For example, at the scene of a burglary or murder, DNA may have been deposited by many people, or transferred there from elsewhere (on a cigarette butt, for example). Matches also include false matches, often because DNA profiles obtained from crime scenes are not complete. For example, the National DNA Database Annual Report 2005–06 states that between May 2001 and April 2006, 50,434 matches with crime scene profiles, or 27.6 percent of the total number of match reports, involved a list of potential suspects, not a single suspect, being given to the police, because matches with multiple records on the NDNAD were made.[34] The report states that this is "largely due to the significant proportion of crime scene profiles that are partial." However, no detailed breakdown of these figures is available and other factors, such as the large number of related individuals who now have profiles on the Database, may also be important.

Table 4.2 shows the number of false matches expected to occur purely by chance between crime scene DNA profiles and the DNA profiles of individuals, calculated by the Forensic Science Service and assuming full

TABLE 4.2 PREDICTED ADVENTITIOUS DNA MATCHES USING FULL PROFILES ON THE NDNAD

Year of	No. on database (millions)	No. of case stains (thousands)	Expected mean no. adventitious matches
2004	2.77	584	2
2005	3.27	634	2
2006	3.77	684	3
2007	4.27	734	4
2008	4.77	784	4
2009	5.27	834	5

Source: National DNA Database Annual Report, 2005–06[a]

profiles are available using Britain's SGM Plus profiling system. This uses ten regions of an individual's DNA (for comparison, the US CODIS system uses thirteen regions, and the old British SGM system uses six). In practice, many more false matches will occur because many crime scene DNA profiles are not complete, although the information that a partial crime scene profile is involved is included in the match report sent to the police.

No details are available on the outcomes of individual match reports as sent to the police. Roughly speaking, for the NDNAD, eight DNA matches lead to four detections, two of which lead to convictions, one of which will involve a custodial sentence.[35] However, only about half of these are *new* detections. In the other cases, the suspect will already have been identified prior to collection of their DNA. These figures are dominated by volume crimes, such as burglaries and theft.

The figures suggest that the majority of individuals who have their name passed to the police in a DNA match report do not subsequently get convicted for an offense. In some cases these individuals may be quickly

a. P. Gill, "National DNA Databases and Some Other Deliberations" (presentation to the Foundation for Science and Technology, February 6, 2008), www.foundation.org.uk/events/pdf/20080206_Gill.pdf.

eliminated from an investigation, without any impact on their lives. But in other cases, even assuming there are no miscarriages of justice, an intrusive investigation may result and the individual may be required to provide alibis or other evidence to demonstrate that they did not commit the alleged offense.

Familial searching is sometimes used when a DNA profile from a crime scene does not match an individual's profile on the Database. Since it is possible that a relative of the suspect is on the Database, looking for a partial match between profiles might identify a parent, child, brother, or sister of the suspect, who can then be interviewed by the police. Familial searching usually produces a long list of names of people to be interviewed and raises ethical concerns because it is possible that it could reveal cases of paternity or nonpaternity that the people interviewed did not know about, and also reveal to relatives who is on the Database. The use of "familial searching" means that anyone who is genetically related to an individual on the Database may also become implicated as a suspect.[36]

New techniques may also increase the risk of false identification and of miscarriages of justice. The increasing use of Low Copy Number (LCN) DNA analysis, which allows a DNA profile to be extracted from a single cell, has led the Director of the Forensic Institute in Edinburgh to warn that innocent people may be wrongly identified as suspects as a consequence of being on the NDNAD[37] and a senior judge to criticize specialist evidence on this technique as contradictory, both during the Northern Ireland Omagh bombing trial and in his detailed judgment in the case.[38] Amongst a string of problems and concerns about contamination, one report claimed that the LCN technique had identified a fourteen-year-old English schoolboy as one suspect for planting the bomb.[39]

LCN analysis and other new techniques such as "DNABoost" increase the sensitivity of DNA analysis (allowing very small samples or mixed samples to be analyzed, respectively) but also increase the chance of a false match between a crime scene DNA sample and an individual's DNA profile.[40] Issues include not only how well the laboratory technique and statistical analysis is validated, but also whether very small DNA samples could be transferred to a crime scene via another person, without the individual ever having been there.[41]

The larger DNA databases become, and the greater the number of comparisons with crime scene DNA profiles, the more false matches are likely to arise. In 2007, the European Union adopted an agreement, based on the Prüm Treaty, to step up cross-border cooperation, particularly in combating terrorism and cross-border crime, which was originally negotiated between a minority of EU countries, led by Germany.[42] A draft agreement on implementation discusses the exchange of DNA data in detail.[43] Although not yet finalized, the draft agreement suggests that matches at only six regions of a person's DNA will be regarded as sufficient for data to be shared. If so, this will considerably increase the likelihood of false matches between crime scene DNA samples collected elsewhere in the European Union and the DNA profiles of individuals with records on the NDNAD. The Information Commissioner warned a House of Lords Committee in 2007[44] that matches based on only six regions of DNA could lead to further cases like that of Raymond Easton, a man who was arrested in England in 1995 (when Britain's DNA profiling system used only six regions) although he could not have been the perpetrator of an alleged crime.[45] The Commissioner also reminded the Committee of a case from 2003 when a UK citizen who had never been to Italy was wrongly arrested for a murder in Italy on the basis of apparent DNA evidence (the case of Peter Hamkin).[46] Hamkin had never been to Italy, but he was arrested on the basis of a DNA match reported by Interpol. He was arrested, taken from Liverpool to London, kept in a police cell overnight, and remained under investigation for twenty days. The police later told him that a "more refined result" from a second DNA sample showed that it was not a match.

FALSE HARASSMENT OR STIGMATIZATION OF MINOR AND YOUNG OFFENDERS

The Database exacerbates the potential for harassment or stigmatization of minor offenders, including racial harassment and the stigmatization of young people who in the past may simply have been given a warning rather than treated as criminals in the making. Other vulnerable people, such as the mentally ill, may also be disproportionately affected. Political protestors can also have their details added and retained on the Database much more easily than in the past. If one person paints a slogan on a wall during a demonstration, for example, a whole group of people may be arrested on suspicion of committing an offense. All will have their DNA and fingerprints taken and retained, even if they did nothing more than join the demonstration.

Matilda MacAttram, Director of Black Mental Health UK, has commented on the implications for black people suffering from mental illness:

> Pathways into care for black patients are invariably via the police or criminal justice system; this means that countless people with healthcare needs are being criminalized in the process of seeking help. It is disturbing to know that those needing healthcare are on a criminal database; wherever this is the case it is imperative that their details are removed as quickly as possible. This begs the question, what kind of a society criminalizes those who need help.[47]

Criminalization may also be particularly significant for children and young people. The system of reprimands and final warnings was developed to keep children out of the courts whenever possible because it was recognized that early exposure to the criminal justice system is counterproductive. Terri Dowty, Director of Action on Rights for Children, argues that the majority of children go through periods of challenging and difficult behavior, and it is important that they can learn from their mistakes and leave difficulties behind upon reaching adulthood.[48] She highlights that significant recent research has demonstrated that contact with the police may have harmful effects for children and young people. The Edinburgh "Study of Youth Transitions and Crime" followed the progress over ten years of four thousand young people who started secondary school in Edinburgh in the autumn of 1998.[49] It found that young people who were caught by the police were more likely to persist in their offending than those who offended at a similar level but who were not caught. This fits with the theory that much youth crime is committed by adolescent offenders who will grow out of crime if they are not damaged by interventions from the criminal justice system.

Retaining DNA profiles and police records can also lead to anxiety for children who have committed no offense, and who may even have been reporting an offense or trying to aid the police with their inquiries. For example, Caitlin Bristow, aged fifteen, was arrested in England in 2005 and had her DNA and fingerprints taken. She had reported an assault and a counterclaim had been made against her, but she was never charged, let alone convicted, of any offense. Caitlin told her local paper, "I'm worried that it will scar my record for life. It might come up if I went for jobs, such as with children—not that I've been in trouble, but just that I'm known to the police."[50]

Focus group research has found that both parents and children have reservations about samples being taken for petty crime and feel that there are dangers in stigmatizing young people for a one-off act.[51] A recent suggestion by the head of forensic sciences at Scotland Yard and the DNA spokesman for the Association of Chief Police Officers (ACPO), Gary Pugh, that children of primary school age should be placed on a national DNA register if they show signs of "becoming a criminal" has also attracted widespread criticism.[52]

TRACKING AND SURVEILLANCE

The rapid expansion of the National DNA Database has enormous implications for the balance between the power of the state to implement "biosurveillance" on an individual and the individual's right to liberty and privacy. There is also significant potential for others, including organized criminals, to infiltrate the system and abuse it, for example by using it to reveal changed identities and breach witness protection schemes.

GeneWatch UK obtained information about the research uses of the DNA Database and samples as a result of a series of Freedom of Information requests made in 2006.[53] The list of projects included some "operational requests," including one on behalf of the police to check the Database for named individuals. One research project involved the selection of some groups of individuals from the Database on the basis of *having African name, having typical Muslim names*, or *having typical Hindu/Sikh names*. In this case, this information was used for research on match probabilities rather than to identify or track the people in these categories. However, it would not be unlawful for the police to use such a process to identify groups of individuals, provided the use of the Database could be claimed to fall within the broad definition of "purposes related to the prevention or detection of crime."

Allowing the Database to be searched by name or using a "familial search" (looking for partial matches between a DNA profile and profiles stored on the Database) means that an individual's DNA profile can be obtained and used to trace their movements or identify relatives. If a person's DNA sample is also accessed, other personal genetic information may also be obtained. The same approach may be used to trace identifiable groups of individuals.

Because an individual may leave DNA wherever they go, there is also potential for it to be used to try to identify whether he or she has been present at scenes other than crime scenes (for example, a political or religious meeting). The legal restriction of uses to "purposes related to the prevention and detection of crime" provides no meaningful barrier to such surveillance, nor is there any independent scrutiny that could identify such uses. Particular concern arises in the context of the right to protest, because acquittal by a court or a spent conviction for a relatively minor offense no longer results in removal of a person's record from the NDNAD or the Police National Computer (PNC).

If criminals can infiltrate the system, they may also be able to use it to track individuals or their relatives. In practice, the process of collecting, analyzing, and storing DNA allows numerous points of access to confidential information (for example, by employees working in the commercial laboratories that analyze and store the DNA samples for the police). If criminals can infiltrate the system, they may be able to use it to identify people whose identity is protected, including people in witness protection schemes and undercover police officers, and to trace their relatives or reveal private genetic information (including paternity and nonpaternity). Vulnerable women and children may be particularly at risk. The risk to privacy is also increased by plans to share more information with EU countries and to check DNA or police records on the spot using handheld devices.[54] A worst case scenario is that someone who infiltrates the law enforcement system of another country, or who gains access to the British system, could use DNA matching to track down a potential victim, by submitting a DNA profile obtained from, say, the toothbrush of a child, rather than a crime scene.

E-mails supplied to GeneWatch UK as a result of a Freedom of Information request in 2006 revealed that the commercial company LGC kept a "minidatabase" of information sent to it by the police, including individuals' demographic details, alongside their DNA profiles and samples.[55] Thus, anonymity is not maintained by separating identifying information from the DNA profiles themselves, because identifying information is sent by the police to the commercial laboratories that analyze the samples.

A 2007 Home Office consultation proposed further extending police powers (outside Scotland) by allowing DNA to be taken on arrest in the street or in short-term holding facilities, in shops or town centers, where people could be detained for up to four hours.[56] Suspected offenses for which DNA can be taken would be expanded to include nonrecordable offenses (such as dropping litter), from anyone aged ten or above. The main purpose of taking DNA and fingerprints would change from investigating offenses to establishing "identity": this implied a new link between the NDNAD and the proposed National Identity Register. The expansion to include nonrecordable offenses was rejected, following opposition from the Association of Chief Police Officers, which warned that "[e]xtending the taking of samples to all offenses may be perceived as indicative of the increasing criminalisation of the generally law-abiding citizen."[57] However, the Counter-Terrorism Act 2008 now allows samples to be used for the purpose of identifying individuals.

DISCRIMINATION BASED ON PERMANENT RECORDS OF ARREST

The retention of permanent records of arrest, on the Police National Computer, is unprecedented in British history. PNC records are available to a wide range of agencies, although a plan is being developed to "step down" records so that access will be limited to the police after similar time frames to those which used to result in their removal.[58] However, information contained in these records may continue to be made available to others as the result of an Enhanced Criminal Record Check. Employers may also lawfully require an individual to undertake his or her own subject access request to the police and reveal this as a condition of employment (known as "enforced subject access"). The retention of permanent records of arrest may have serious potential consequences for an individual, including refusal of visas or access to visa waiver schemes, refusal of employment in any occupation not covered by the Rehabilitation of Offenders Act (including all jobs working with children or young people, and a wide range of professions, such as the legal profession), and excessive government or police surveillance (of individuals or selected groups of people). UK citizens who have been arrested (whether they have been convicted or not) are now ineligible for the US Visa Waiver scheme and must now undergo the expensive and time-consuming process of applying for a full visa to travel to the United States. As part of this process, they are required to pay for the police to release their record of arrest to the US embassy.

GENETIC RESEARCH WITHOUT CONSENT

The NDNAD has routinely been used for research without consent. In March 2005, the Home Office was severely criticized by the House of Commons Science and Technology Committee for allowing research without consent or any ethical oversight. An ethics committee was established and held its first meeting in October 2007. Although new procedures for approving research projects are being adopted, it remains a matter of concern that controversial genetic research may take place without consent.

Freedom of Information requests made by GeneWatch UK in 2006 showed that since the year 2000, nineteen research projects had been allowed and fourteen refused.[59] The requests revealed that stored DNA samples have been used for genetic studies of the male Y chromosome without the consent of the people involved as part of a controversial attempt to predict ethnicity from DNA. This type of research could also inadvertently reveal other genetic characteristics such as a man's risk of infertility. Despite numerous requests for information, the list of research projects is still incomplete.

Retention of individuals' DNA samples increases privacy concerns, and the new legal requirement to destroy them has been widely welcomed. Such samples are already destroyed in some other countries, such as Germany, once the computerized DNA profiles used for identification purposes have been obtained. The Home Office has recognized that retaining samples is "one of the most sensitive issues to the wider public,"[60] and the Human Genetics Commission concluded that the reasons given for retaining them are "not compelling."[61] Only temporary, not permanent, storage is necessary for quality assurance purposes and a new sample can always be taken from the suspect if a DNA profile requires checking or upgrading.

The research that has been done using the Database with the aim of trying to predict an individual's ethnic appearance from their DNA is particularly controversial. It is part of the research and development of a new commercial product: a DNA test to predict ethnicity or ancestry. This research has used both the DNA profiles on the computer Database and the stored DNA samples. The expected role in criminal investigations of predicting ethnicity or ancestry from DNA is to try to build up a "genetic photo fit" of a suspect purely from a crime scene DNA sample.

Historically, genetic explanations of race have been used against ethnic minority groups, causing stigma and discrimination, and being used to justify racism, colonialism, and eugenics.[62] More recent research suggests that there is a complicated relationship between genetic differences and what is commonly called "race." Human beings are all one species and biologically distinct races do not exist.[63] The relationship between skin color and ancestry is also complex[64] and also appears to have been influenced by social factors (the racist treatment of people identified as black).[65]

To some extent broad geographical ancestry (for example, Africa, Europe, or Asia) can be predicted from the frequency of different genes.[66] Many companies are now selling genetic ancestry tests commercially: some market these tests to the police as well as individuals. There are two main techniques:[67]

(1) *Lineage-based tests*, which try to trace the inheritance of some of a person's DNA through the male or female line. Maternal lineage-based tests use mitochondrial DNA (mtDNA), which a person inherits only from their mother. Paternal lineage-based tests use the Y chromosome, which men inherit from their father. Some relatively rare lineages can be traced to particular ethnic groups or locations using this method, but others are much harder to place. Predictions in mixed urban populations are likely to be much less reliable.[68]

(2) *Bio-geographical ancestry tests* use the statistical distribution of different genetic markers in different countries and different ethnic groups. These tests try to estimate a person's ancestry from the percentage of

> these markers that they have. However, the results are of questionable reliability because they depend on the regions considered, the number of genes tested, and the extent to which populations have mixed in the past.

The Nuffield Council on Bioethics has warned, "In view of the significant ethical and practical problems, and the limited usefulness of the information provided, attempts to infer ethnicity from DNA profiles and samples fail the test of proportionality and we recommend that ethnic inferences should not be routinely sought, and should be used with great caution." However, people with DNA profiles on the Database cannot refuse to give consent to their DNA profiles being used for this type of research.

The Database appears not to have been used for behavioral genetic research. However, this type of research is equally—if not more—controversial. It could also be conducted under current legislation without the consent of the people whose genetic data is used. A 2007 study assessed the views of criminal justice practitioners about behavioral genetics.[69] The study included barristers, solicitors, judges, probation officers, and social workers who are involved in the management of individuals that may be deemed at risk of displaying violent and aggressive behaviors. It identified concerns that a policy emphasis on aggressive and antisocial behavior exacerbates the shift toward further control and surveillance of citizens, particularly of those deemed "risky," who are already overpatrolled. It is possible that genetic information relating to behavior could be slotted into existing systems of profiling and collating information on individuals, including children. In June 2008, a committee of MPs warned against profiling to predict criminal behavior and stated that it would be particularly concerned if information held on children was used for the purposes of predictive profiling rather than child protection.[70] The reliability and predictive value of such information is extremely questionable.

DISCUSSION AND CONCLUSIONS

GeneWatch UK recognizes the extremely important role that DNA can play in some criminal investigations. We are not opposed to the existence of the National DNA Database, but are deeply concerned that its rapid expansion has spiraled out of control. The law in England, Wales, and Northern Ireland allows the capture and use of genetic information without consent from a defined section of the community (those who have been arrested for a recordable offense), often referred to as the "active criminal population," despite the fact that many of these individuals have not committed any crime. There is a strong bias in the system toward the inclusion of DNA profiles from young black men and vulnerable people, including children and the mentally ill.

The rapid expansion of the Database has enormous implications for the balance between the power of the state to implement "biosurveillance" on an individual and the individual's right to privacy. There is also significant potential for others—including organized criminals—to infiltrate the system and abuse it, for example by using it to reveal changed identities and breach witness protection schemes.

The permanent retention of all police records significantly changes the relationship between the individual and the state. Individuals with records on the DNA Database lose their presumed legitimacy to go about their daily life and their right to keep their family relationships and some other genetic information private. Even if they have never been charged or convicted of any offense, they may be refused employment or a visa as a result of the retention of a permanent record of their arrest on the Police National Computer (PNC). The retention of an individual's DNA profile also allows their movements to be tracked or their relatives to be identified. The potential implications for the right to protest are particularly serious.

There are many circumstances in which the retention of an individual's DNA profile and linked data will give rise to potential identification—only in the minority, not the majority, of cases does this involve the identification of the individual as the perpetrator of a crime. Many individuals identified through matches on the Database will be subject to investigation by the police, but are subsequently acquitted of any crime. The purpose of data retention is quite different from the purpose of collection, since it is a form of surveillance based on the idea that the individual, or a relative of theirs, may commit a *future* crime, not that they have already committed one.

It is difficult to reconcile the current situation with the principle of equal application of the law (the concept that everyone is equal before the law). People on the Database are treated as members of a "risky population" whose DNA requires retention by the State.[71] Young people, people suffering from mental illness, and people from black and minority ethnic groups are particularly likely to be members of this "risky population." Retention of an individual's DNA profile on a Database is likely to be of most benefit when he or she has a record as a "career criminal" and is considered likely to reoffend. However, the population on the Database now includes anyone who is arrested for a recordable offense.

The lesson from the rapid expansion of the National DNA Database is that there is significant potential for stigma and prejudicial treatment of people who have their DNA profiles retained. Although putting everyone on the Database is sometimes proposed as a solution to discrimination, it would not prevent the Database being used in a discriminatory way and would considerably exacerbate concerns about potential misuse and about false matches. Such proposals are also widely regarded as extremely costly and impractical.

The rapid expansion of the National DNA Database has not improved the crime detection rate.[72] GeneWatch UK believes that there are important changes that could be made that would improve safeguards for human rights and privacy without compromising the role of the DNA Database in tackling crime. Most people should have their records removed immediately if they are not convicted, with a possible temporary exception for some people accused of serious or violent offenses, as is the case in Scotland. Restrictions on retention should apply to police computer records, as well as DNA profiles and fingerprints. All DNA samples should be destroyed once the computerized DNA profiles needed for identification purposes have been obtained from them. People with convictions or cautions for minor crimes should not have their records retained indefinitely. A more restrictive policy for sampling DNA, especially for juveniles, also needs to be adopted.

NOTES

The author thanks Terri Dowty, Director of Action on Rights for Children (www.arch-ed.org), for providing the information regarding the arrest of children and young people.

1. R. Williams and P. Johnson, "Circuits of Surveillance," *Surveillance and Society* 2, no. 1: 1–14.

2. Parliamentary Question, House of Commons Hansard, October 30, 2007: Column 1254W.
3. Parliamentary Question, House of Commons Hansard, February 22, 2010: Column 350W.
4. R. Williams, P. Johnson, and P. Martin. "Genetic Information and Crime Investigation" (August 2004). The Wellcome Trust, www.dur.ac.uk/p.j.johnson/Williams_Johnson_Martin_NDNAD_report_2004.pdf.
5. FSNI-PSNI Service Level Agreement 2006–2007.
6. Her Majesty's Inspectorate of Constabulary, "Under the Microscope," July 1, 2000, http://inspectorates.homeoffice.gov.uk/hmic/inspect_reports1/thematic-inspections/utm001.pdf.
7. www.publications.parliament.uk/pa/cm200001/cmhansrd/vo010314/debtext/10314-32.htm.
8. The Criminal Justice (Northern Ireland) Order 2004, Statutory Instrument 2004 No. 1500 (N.I. 9), www.opsi.gov.uk/si/si2004/20041500.htm.
9. GeneWatch UK, "Using the Police National DNA Database—Under Adequate Control?" GeneWatch Briefing, June 2006, www.genewatch.org.
10. Council of Europe, Recommendation No. 92 on the use of analysis of deoxibonucleic acid (DNA) within the framework of the criminal justice system (adopted on 10 February 1992).
11. Nuffield Council on Bioethics, "The Forensic Use of Bioinformation: Ethical Issues," Nuffield Council in Bioethics, September 2007, www.nuffieldbioethics.org/go/ourwork/bioinformationuse/introduction.
12. Case Of S. And Marper v. The United Kingdom. Grand Chamber of the European Court of Human Rights. Judgment. Strasbourg, December 4, 2008.
13. Crime and Security Act 2010, www.opsi.gov.uk/acts/acts2010/pdf/ukpga_20100017_en.pdf.
14. Scottish Parliament Justice 2 Committee Official Report, March 28, 2006, www.scottish.parliament.uk/business/committees/justice2/or-06/j206-0902.htm#Col2146; Scottish Parliament Official Report. Police, Public Order and Criminal Justice (Scotland) Bill: Stage 3. May 25, 2006. www.scottish.parliament.uk/business/official Reports/meetingsParliament/or-06/sor0525-01.htm.
15. www.scotland.gov.uk/News/Releases/2007/01/29133555.
16. Home Office Circular 16/95.
17. F. Coates, "Police to File All Offences for Life," *Times*, January 21, 2006, www.timesonline.co.uk/section/0,2086,00.html
18. House of Commons Hansard, October 9, 2006: Column 493W.
19. House of Commons Hansard, February 9, 2010: Column 909W.
20. Home Office, "DNA Expansion Programme 2000–2005: Reporting Achievement," Forensic Science and Pathology Unit, http://police.homeoffice.gov.uk/news-and-publications/publication/operational-policing/DNAExpansion.pdf.
21. National DNA Database. Adjournment Debate. House of Commons Hansard 29 Feb 2008: Column 1427, www.publications.parliament.uk/pa/cm200708/emhansard/cm080229/debtext/80229-0013.htm.

22. B. Leapman, "Three in Four Young Black Men on the DNA Database," *Telegraph*, November 5, 2006, www.telegraph.co.uk/news/uknews/1533295/Three-in-four-young-black-men-on-the-DNA-database.html.
23. House of Commons Home Affairs Committee, "Young Black People and the Criminal Justice System," Second report of session 2006/07, Volume I, HC-181-I, www.publications.parliament.uk/pa/cm200607/cmselect/cmhaff/181/181i.pdf.
24. C. Sharp and T. Budd, "Minority Ethnic groups and Crime: Findings from the Offending, Crime and Justice Survey 2003," Home Office Online Report 33/05, www.homeoffice.gov.uk/rds/pdfs05/rdsolr3305.pdf
25. House of Commons Hansard, May 10, 2007, Col.430W.
26. BBC News, "Police Condemn 'Target Culture,'" May 2007, http://news.bbc.co.uk/1/hi/uk/6656411.stm.
27. S. Hoggart, "Targeting the Young," *Guardian*, March 1, 2008, www.guardian.co.uk/politics/2008/mar/01/1.
28. "Wasting Police Time," BBC Panorama, September 16, 2007, http://news.bbc.co.uk/1/hi/programmes/panorama/6948201.stm.
29. Uncorrected transcript of oral evidence. To be published as HC 364-ii, www.publications.parliament.uk/pa/cm200708/cmselect/cmhaff/uc364-ii/uc36402.htm.
30. "A Temporary Respite," Guardian, February 2007, www.guardian.co.uk/commentisfree/2007/feb/19/comment.politics3.
31. Home Office Statistical Bulletins, "Arrests for Recorded Crime etc.," 02/03–05/06.
32. Youth Justice Board Annual Statistics, 03/04–05/06.
33. GeneWatch UK and ARCH, "How Many Innocent Children Are on the National DNA Database?" www.genewatch.org/uploads/f03c6d66a9b354535738483c1c3d49e4/Childrenfigsbrieffinal.doc.
34. National DNA Database Annual Report 2005–06, www.homeoffice.gov.uk/documents/DNA-report2005-06.pdf.
35. GeneWatch UK, "The DNA Expansion Programme: Reporting Real Achievement?" February 2006, www.genewatch.org/uploads/f03c6d66a9b354535738483c1c3d49e4/DNAexpansion_brief_final.pdf.
36. R. Williams and P. Johnson, "Inclusiveness, Effectiveness and Intrusiveness: Issues in the Developing Uses of DNA Profiling in Support of Criminal Investigations," *Journal of Law and Medical Ethics* 33, no. 3: 545–58.
37. J. Morgan, "Guilty by a Handshake?" *Herald*, May 2, 2006.
38. "Fresh Criticism of Omagh Evidence," *BBC Online*. December 8, 2006, http://news.bbc.co.uk/1/hi/northern_ireland/6162483.stm; "The Queen v Sean Hoey," 21 December 2007, http://business.timesonline.co.uk/tol/business/law/article3083217.ece.
39. B. McCaffrey, "Controversial DNA Tests Identified Schoolboy as Part of Omagh Attack," *The Sunday Business Post*. November 12, 2006, http://archives.tcm.ie/businesspost/2006/11/12/story18791.asp.
40. "New DNA Test to Solve More Cases," *BBC Online*, October 4, 2006, http://news.bbc.co.uk/1/hi/england/5404402.stm.

41. A. Jamieson "Mixed Results," *Guardian*, February 28, 2008, www.guardian.co.uk/commentisfree/2008/feb/28/ukcrime.forensicscience.
42. www.statewatch.org/news/2008/feb/eu-prum-decision.pdf.
43. www.statewatch.org/news/2008/feb/eu-prum-implementing-draft.pdf.
44. House of Lords Select Committee On The European Union. Sub Committee F (Home Affairs). Inquiry into the Prüm Convention. Evidence submitted by the Information Commissioner.
45. "Disabled Man Turns Down Payout Offer," *Swindon Advertiser*, August 15, 2000. http://archive.thisiswiltshire.co.uk/2000/8/15/238098.html.
46. C. Johnson, "Cleared Murder Accused Victim of DNA Blunder," *Liverpool Daily Post*, March 10, 2003, http://icliverpool.icnetwork.co.uk/0100news/0100regionalnews/page.cfm?objectid=12718961&method=full&siteid=50061.
47. "Calls Mount for Removal of Mental Health Patients from DNA Dataset on Eve of House of Commons Debate," Black Mental Health. February 28, 2008, www.blackmentalhealth.org.uk/index.php?option=com_content&task=view&id=266&Itemid=150.
48. T. Dowty and H. M.Wallace, "DNA Database: Fuelling Children's Criminality?" ChildRight, May 2008.
49. Edinburgh Study of Youth Transitions and Crime, www.law.ed.ac.uk/cls/esytc/.
50. "Dad Vows 'Fight Will Go on for Daughter's DNA,'" *The Wilmslow Express*, January 25, 2006.
51. M. Levitt and F. Tomasini, "Bar-coded Children: an Exploration of Issues Around the Inclusion of Children on the England and Wales National DNA database," *Genomics, Society and Policy*, 2, no. 1: 41–56, www.gspjournal.com.
52. E. James, "DNA Initiative Brands Children as Criminals," *Guardian*, March 25, 2008, www.guardian.co.uk/society/2008/mar/25/children.youngpeople.
53. GeneWatch UK, "Using the Police National DNA Database—Under Adequate Control?" GeneWatch Briefing, June 2006, www.genewatch.org.
54. L. Adams, "Police Computer Goes on the Beat," *Herald*, October 14, 2006, www.theherald.co.uk/news/72189.html; for example, www.itweek.co.uk/vnunet/news/2170113/portable-dna-analyzer-invented.
55. www.genewatch.org/uploads/f03c6d66a9b354535738483c1c3d49e4/AnswerFOI8May.pdf; A. Barnett, "Police DNA Database is 'Spiraling Out of Control,'" *Observer*, July 16, 2006, http://observer.guardian.co.uk/uk_news/story/0,1821676,00.html.
56. Home Office, "Modernising Police Powers," Review of the Police and Criminal Evidence Act (PACE) 1984. Consultation Paper. Home Office, March 2007.
57. Home Office, Pace Review: Summary of Responses to Public Consultation 16 March—31 May 2007, http://police.homeoffice.gov.uk/publications/operational-policing/PACEReviewRespondentsSummar1.pdf?view=Binary.
58. ACPO, Retention guidelines for nominal records on the Police National Computer, March 16, 2006.
59. GeneWatch UK, "Using the Police."

60. Home Office, Supplementary Memorandum, Appendix 20. In: House of Commons Science and Technology Committee, vol. 2, *Forensic Science on Trial*. HC 96-II, www.publications.parliament.uk/pa/cm200405/cmselect/cmsctech/96/96ii.pdf.
61. Human Genetics Commission, "Inside Information," May 2002, www.hgc.gov.uk/UploadDocs/DocPub/Document/insideinformation_summary.pdf; Human Genetics Commission, "HGC Response to the Scottish Executive Consultation on Police Retention of Prints and Samples," www.scotland.gov.uk/Resource/Doc/77843/0018244.pdf.
62. H. Bradby, "Genetics and Racism," in *The Troubled Helix*, ed., T. Marteau, M. Richards (Cambridge: Cambridge University Press 1996).
63. S. O. Y. Keita, R. A. Kittles , C. D. M. Royal et al., "Conceptualizing Human Genetic Variation," *Nature Genetics Supplement* 36, no. 11 (2004): S17–S20.
64. M. D. Shriver, E. J. Parra, S. Dios et al., "Skin Pigmentation, Biogeographical Ancestry and Admixture Mapping," *Human Genetics* 112 (2003): 387–99; E. J. Parra, R. A. Kittles and M. D.Shriver, "Implications of Correlations Between Skin Color and Genetic Ancestry for Biomedical Research," *Nature Genetics Supplement* 36, no. 11: S54–S60.
65. F. W. Sweet, "Afro-European Genetic Admixture in the United States," June 8, 2004, http://backintime.com/Essay040608.htm.
66. M. Bamshad, S.Wooding, B. A. Salisbury et al., "Deconstructing the Relationship Between Genetics and Race," *Nature Reviews Genetics* 5: 598–609.
67. M. D. Shriver and R. A.Kittles, "Genetic Ancestry and the Search for Personalized Genetic Histories," *Nature Reviews Genetics* 5 (2004): 611–18.
68. M. A. Jobling, "Y-chromosomal SNP Haplotype Diversity in Forensic Analysis," *Forensic Science International* 118: 158–62.
69. E. Pieri and M. Levitt, "Genetics and Crime: Policy and Practice Impacts of Behavioural Genetics Research into Aggressiveness and Violence," http://onlinelibrary.wiley.com/doi/10.1111/j.1467-8519.2008.00694.x/full.
70. House of Commons Home Affairs Committee, "A Surveillance Society?" Fifth report of Session 2007–08, vol. 1, www.publications.parliament.uk/pa/cm200708/cmselect/cmhaff/58/58i.pdf.
71. C. McCartney, "Forensic DNA Sampling and the England and Wales National DNA Database: a Sceptical Approach," *Critical Criminology* 12: 157–78.
72. GeneWatch UK, "Submission to the Home Affairs Committee: The National DNA Database," January 2010, www.genewatch.org/uploads/f03c6d66a9b354535738483c1c3d49e4/GWsub_Jan10.doc.

PART III
ANCESTRY TESTING

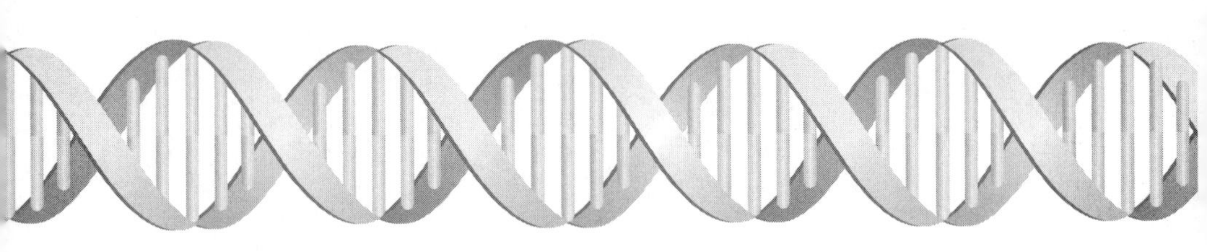

5
ANCESTRY TESTING AND DNA

USES, LIMITS, AND *CAVEAT EMPTOR*

Troy Duster

Direct consumer use of DNA tests for ancestry tracing has taken off in the last five years, and we are not just talking about probes for first-generation genetic lineage as in "Who's your daddy?" popularized on daytime "reality" television. Since 2002, nearly a half-million people have purchased tests from at least two dozen companies marketing direct-to-consumer kits.[1] The motives for testing range from the desire for ancestral links to those who lived on other continents over five hundred years ago, to a more modest interest in reconstructing family histories. For many African Americans, the quest to find a link to regions and peoples of sub-Saharan Africa can take on a spiritual or even messianic quest, at least partially explained by the fact that the Middle Passage across the Atlantic during the slave trade explicitly and purposefully obliterated linguistic, cultural, religious, political, and kinship ties. The 2006 PBS television series *African American Lives* brought this quest into sharp relief. First celebrity and later ordinary blacks were mesmerized by stories of DNA matches that claimed to reveal or refute specific ancestral links to Africa, Native American heritage, and, surprising to some, East Asian or European populations.

In sharp contrast, CBS's *60 Minutes* aired a dramatic segment on October 7, 2007, that portrayed a direct and sharp challenge to the claims made about such ancestry testing. The segment began with Vy Higgensen, an

African American woman from New York's Harlem triumphantly affirming her connection to "new kin" (one of whom was a white male cattle rancher from Missouri). But as the program unfolds, we see a disturbing cloud of doubt drift over the last part of the segment, which ends with a less than subtle hint that the claims are specious. A first test from the company African Ancestry claims that Higgensen is linked to ancestors in the Sierra Leone, the Mende people. She rejoices, "I am thrilled! It puts a name, a place, a location, a people!" But then she is shown the results of a second test, from another company, Relative Genetics, which claims that she instead has a genetic match to the Wobe tribe of the Ivory Coast. She seems philosophical. Yet a third test, from still another company, Trace Genetics, claims that her ancestors are from Senegal, the Mendenka. Now she seems agitated, visibly concerned, confused, and most certainly disappointed that what began as a definitive match to a particular group or region of Africa has now turned into a "you pick which one you want to believe" game.

The very next month, serious questions about the tests were revisited when Henry Louis Gates, who had hosted the aforementioned *African American Lives*, said that the same thing had happened to him. Here is how the *New York Times* (November 25, 2007) cast the story:

> Henry Louis Gates Jr., whose PBS special "African American Lives" explores the ancestry of famous African-Americans using DNA testing, has done more than anyone to help popularize such tests and companies that offer them. But recently this Harvard professor has become one of the industry's critics.
>
> Mr. Gates says his concerns date back to 2000, when a company told him his maternal ancestry could most likely be traced back to Egypt, probably to the Nubian ethnic group. Five years later, however, a test by a second company startled him. It concluded that his maternal ancestors were not Nubian or even African, but most likely European.
>
> Why the completely different results? Mr. Gates said the first company never told him he had multiple genetic matches, most of them in Europe. "They told me what they thought I wanted to hear," Mr. Gates said.

Here we have the first sally into a combined definitional and epistemological conundrum, beginning with the meaning of "ancestry." While this is typically used to refer to geographic areas where one's biological ancestors lived, with just a few minutes of reflection, we can see an enormous problem to which even common sense will alert us: which ancestors? Easy enough if we are only dealing with mom and dad, or 4 grandparents—we can even handle three generations back with 8 great-grandparents. But if we go back six generations, that means we all have 64 direct biological ancestors. Since each of these 64 could be said to have made an equal biological contribution to our makeup, why would we choose to represent any one or two as our "real" biological lineage? (Eight generations gives us 256 such ancestors, and twenty generations places the figure at 1,048,576.)

THE CAPACITIES AND LIMITS OF USING DNA TO TEST FOR ANCESTRY

What can DNA tell us about our genetic lineage, and where does it fall short? What explains Vy Higgensen's multiple results from different testing sites? Flawed methodology? Partial truths hyped as definitive findings? Did the testing companies use different methods, or deploy different reference populations, or both?

Let's begin with what DNA testing *can* tell us about biological ancestry. There are two different tests—one for males and another for females—and each can provide relatively definitive results along one particular line of our genetic ancestry.

Males inherit the Y Chromosome from their biological fathers. The markers are sufficiently distinctive so that the test can not only identify the father, but also the father's father, and if the data were available, the father's father's father. This path to ancestry identification can go on for as many generations as data are available for—which is how Thomas Jefferson (or one of his brothers) was linked to Sally Hemings's offspring. For more than 150 years, historians argued and debated as to whether Jefferson had children with one of his slaves, Sally Hemings. Only in the last decade has Y Chromosome analysis settled the debate in favor of those who have claimed that the historical record pointed to Thomas Jefferson.

The test for female ancestry has an interesting parallel. We can definitively answer "Who's your mommy?" All of a mother's children inherit her mitochondrial DNA (mtDNA). Located within the cell but outside the nucleus, mtDNA serves as the cell's energy producers but only the daughters pass it on. Thus, for a female, it is possible to trace and identify her mother, her mother's mother, etc., (along the same line as just noted for males using Y chromosome analysis). This was the way that granddaughters were linked to their grandmothers in the aftermath of Argentina's Dirty War (1976–83). Thousands of young fathers and mothers "disappeared" by acts of the ruling junta, and their orphaned small children were given to couples who wished to adopt.[2] It was through mitochondrial DNA testing that the grandmothers were reunited with the children of their daughters (who had been murdered or had disappeared). These two tales reveal not only the power of DNA ancestry testing, but their significant and consequential social and political uses as well.

But it is also vital to restate the limitations, that is, that these two tests can identify, for example, only 2 of the 64 great-great-great-great-grandparents. Indeed, only 2 of the next generation back, of 128, can be so identified, only 2 of 256, and so on. Yet each of the other 62 or 126 or 254 contributed as much to our genetic makeup as the two we can trace by the sex-linked paternal or maternal lines. The Genographic Project of the *National Geographic Magazine* uses these two tests, supplemented by a selection of twenty-two additional markers. The researchers correctly inform participants who send in their DNA that there are limitations to what can be claimed. Nonetheless, people who receive the results are often led to believe that if their test does *not* match the archival sample of a particular Native American or indigenous group in Canada and Alaska, then they are *not* genetically linked to that group. Several years ago, when Genographic scientists sampled people in the Arctic North, Lorianne Rawson, a forty-two-year-old woman who had strong social ties to, and who believed that she was descended from, the Aleuts of Alaska, submitted her DNA to the Genographic project. She was informed by the testers that results linked her instead to the Yup'ik Eskimos, the enemies of the Aleuts.[3] Personal and political trauma can understandably ensue from such seemingly authoritative reassignments. This kind of "result," however problematic in terms of disclaimers or caveats, happens when the technology inevitably limits the analysis to particular corridors or silos of the ancestry tree, and locks in on that limited corridor. While the results are presented as an authoritative claim, the laity are not provided with the tools to understand how the many other ancestral links noted above are excluded by the limits of ancestry tracing through DNA analysis.

Sometimes these putative links (or lack of them) have significant financial repercussions. The black Seminoles have been struggling with this very question of whether to use DNA analysis to "authenticate" their relationship to the Seminole Indian Tribe. The reason is straightforward and serious: money. The federal government, pursuant to a land-settlement claim, made an award to Seminole Indians in 1976, poised to distribute upward of $60 million. In 2000, the Seminole Nation of Oklahoma amended its constitution so that members needed to show "one-eighth Seminole blood."[4] The black Seminoles could use either Y chromosome analysis or mitochondrial DNA (mtDNA) to link themselves through very thin chains back along two edges of the genealogical axis (mother's mother's mother, etc.; or father's father's father, etc.), but that would miss all other grandparents (fourteen of sixteen, thirty of thirty-two, sixty-two of sixty-four). The stakes are even higher for the Florida Seminoles. In 2006, the tribe purchased the entire Hard Rock Café chain for approximately $1 billion. If you were offered a genetic ancestry test of either Y Chromosome or mtDNA analysis, would you really want to engage the probabilistic Russian-roulette-type gamble?

To supplement the limitation of Y Chromosome and Mitochondrial DNA testing, a group of researchers has come up with a procedure to discern the frequency of certain markers that are hypothesized as belonging, selectively, to our ancestors. However, there are several blind assumptions that have to be accepted in order to have confidence in the links to ancestral populations so defined.

ANCESTRAL INFORMATIVE MARKERS (AIMS)— THE NEW PROXY FOR RACE

Unlike Y chromosome or mtDNA tests, this technology examines a group's relative share of genetic markers found on the autosomes—the nongender chromosomes inherited from both parents. Since Ancestry Information Markers (AIMs) are overwhelmingly shared across all human groups, it is therefore not their absolute presence or absence, but their rate of incidence, or frequency, that is usually being analyzed, and this is especially true when it comes to claims about continental populations. How did these markers come to represent ancestral populations of Africa, Europe, and Native America? The vast majority of these markers are *not* "population specific," as the inventor of Ancestry Informative Markers (AIMs) originally claimed.[5] Because the companies marketing ancestry tests hold proprietary interests in their techniques, most do not make them available for possible scientific replication, and their modeling constructs are therefore undisclosed. Thus, we are left to speculate about the threshold level of frequency that is used to determine the grounds for inclusion or exclusion, as well as what counts as a "pure" referent population.

In one lab that permitted its procedures to be studied by a medical anthropologist, ancestry percentages were generated by formulas that compare the relative frequency of markers (forty-four in total) between selected populations of recent European, African, and Native American descent.[6] All those in the defined group were tested for the frequency of markers that the researchers hoped would provide *relative* distinguishability. Recall that the frequency at which each marker appears in each group is noted—and whole continents are never sampled. Finally, the researchers compare marker frequencies between the three groups to come up with values that, when taken together, yield a probability result about ancestral percentages. This procedure generates the baseline for the statistically based notion of a 100 percent pure European (or African, etc.), so that when you send in your DNA from the saliva swab, and it turns out that you have one-third of the markers that have been designated as "European"—you are told that you are 33 percent European. It is by this statistical legerdemain that we have come to the molecular reinscription of race in contemporary human genetics.[7]

There are a number of deeply problematic, even flawed assumptions behind that percentage claim. What is this "reference population" that has become the measuring stick by which we inform people of their "percent ancestry to a putatively pure continental population" (read "race" here). Let's reexamine such a result if reported back to someone of recent African descent. First, more than seven hundred million people currently inhabit the African Continent—and human geneticists have known for decades that this is the continent with the greatest amount of genetic variation on the globe. The reason for this variation was noted by Pilar

Ossorio: "For many regions of the human genome, there are more variants found among people of Africa than found among people in the rest of the world. This is probably because humans have resided in Africa for much longer than we have resided any place else in the world, so our species had time to accumulate genetic changes within the people in Africa."[8]

A scientifically valid random sampling of even 1 percent of this population would require a prohibitively expensive research program, a database of seven million. So instead, researchers have settled for "opportunity samples," namely, a few hundred here or there, or even thousands that have been collected for a variety of reasons. No attempt has ever been made to take theoretically driven or random samples from African tribes such as the Lua, Kikiyu, Ibo, Hauser, Bantu, Zulu (with all the linguistic, cultural and political complexities of defining the boundaries of such groups), not to mention the thousands of language groups spread across the continent. How then, can we have any sense of reliability or validity for a claim that says someone is 80 percent African—when the baseline for that claim is based on the transparent scaffolding of chance—not purposive sampling?

Yet, when taken together, we are told that these markers appear to yield sufficiently distinctive patterns in those continental populations tested. So now we see how a specific pattern of genetic markers on each of a set of chromosomes that have a *higher frequency* in the "Native Americans" sampled becomes established as a "Native American" ancestry reference. (The fact that there are more than 480 different populations of the Tribal Council,[9] of which the vast majority have never been sampled, is no small matter here, but that is not the focus of the critique I am about to make.) The problem is that millions of people around the globe will have a similar pattern. That is, they share similar base-pair changes at the genomic points under scrutiny. This means that someone from Bulgaria whose ancestors go back to the fifteenth century could (and sometimes does) map as partly "Native American," although no direct ancestry is responsible for the shared genetic material. There is an overwhelming tendency for those who do AIMs analysis with the purpose of claims about ancestry to arbitrarily reduce all such possibilities of shared genotypes to "inherited direct ancestry." In so doing, the process relies excessively on the idea of 100-percent purity, a condition that could never have existed in human populations.

While this is a huge problem, yet another issue looms even larger. If a computer program produces an outcome indicating that 35 percent or more of a particular genetic marker exists in population A (let's call them East Asian), while 35 percent or less occur in population B (let's call them European), the researcher may use that marker to say that someone is from East Asian ancestry. To make matters even more complicated, claims about how a test subject's patterns of genetic variation map to continents of origin and to populations where particular genetic variants arose require that the researchers have "reference populations." The public needs to understand that these reference populations comprise relatively small groups of *contemporary* people. Those groups sampled may have migrated over several centuries, and thus these researchers must make many untested assumptions in using these contemporary groups to stand as proxies for populations from centuries ago, whether putatively representing a continent, a region, or a linguistic, ethnic, or tribal group. To construct tractable mathematical models and computer programs, researchers bracket these assumptions about ancient migrations, reproductive practices, and the demographic effects of historical events such as plagues and famines. Given these intractable barriers to even low-level probabilistic reliability, geneticists are on thin ice telling people that they do or do not have ancestors from a particular people.

Thus, instead of asserting that someone has no Native American ancestry, the most truthful statement would be: *It is possible that while the Native American groups we sampled did not share your pattern of markers, others might since these markers do not exclusively belong to any one group of our existing racial, ethnic, linguistic, or tribal typologies.* But computer-generated data provide an *appearance* of precision that is dangerously seductive and equally misleading. Now we come to one part of the answer as to why different companies come to different results. We cannot conclude that an individual has a close affinity to a particular ethnic or racial group or local geographical population simply because their DNA markers match that population: "Such a conclusion would require demonstrating that the DNA sequence is not present in other places, it would require demonstrating that the gene pool of that ethnic group or local population had been close and immobile for centuries and millennia."[10]

BE ESPECIALLY WARY OF APPLICATIONS OF THESE CLAIMS

There is a yet more ominous and troubling element of the reliance upon DNA analysis to determine who we are in terms of lineage, identity, and identification. The very technology that tells us what proportion of our ancestry can be linked, proportionately, to sub-Saharan Africa (ancestry informative markers) is the same being offered to police stations around the country to "predict" or "estimate" whether the DNA left at a crime scene belongs to a white or black person. This "ethnic estimation" using DNA relies on a social definition of the phenotype (phenotype being the observable physical or biochemical characteristics of an organism, determined by both genetic makeup and environmental influences). That is, to say that someone is 85 percent African, we must know who is 100 percent African. Any molecular, population, or behavioral geneticist who uses the term "percent European" or "percent Native American" is obliged to disclose that the measuring point of this "purity" (100 percent) is a statistical artifact that begins not with the DNA, but with a researcher's adopting the folk categories of race and ethnicity.

THE SEGUE TO FORENSICS AND CRIMINAL JUSTICE AND "MOLECULAR RACE"

It is possible to make arbitrary groupings of populations (geographic, linguistic, self-identified by faith, identified by others by physiognomy, etc.) and still find statistically significant genetic markers between those groupings. For example, we could simply pick all of the people in Chicago and Los Angeles and find statistically significant differences in DNA marker frequency at *some* loci. Of course, at many loci, even most loci, we would not find statistically significant differences. When researchers claim to be able to assign people to groups based on marker frequency at a certain number of loci, they have chosen loci that show differences between the groups they are trying to distinguish.

The work of Evett et al., Lowe et al., and others suggest that there are only about 10 percent of sites in the DNA that are "useful" for making distinctions.[11] This means that at the other 90 percent of the sites, the allele (one member of a pair or series of genes that occupy a specific position on a specific chromosome) frequencies do not vary between groups such as "Afro-Caribbean people in England" and "Scottish people in England." But it does not follow that because we cannot find a single site where allele frequency matches some phenotype that we are trying to identify (for forensic purposes, we should be reminded), that there are not several (four, six, seven) that will not be effective, for the purposes of aiding the FBI, Scotland Yard, or the criminal justice systems around the globe in highly probabilistic statements about suspects, and the likely ethnic, racial, or cultural populations from which they can be identified statistically.

So when it comes to molecular biologists asserting that "race has no validity as a scientific concept," there is an apparent contradiction with the practical applicability of research on allele frequencies in specific populations. It is possible to sort out and make sense of this, and even to explain and resolve the apparent contradiction, but only if we keep in mind the difference between using a taxonomic system with sharp, discrete, definitively bounded categories, and one which shows patterns (with some overlap), but which may prove to be empirically or practically useful.

When representative spokespersons from the biological sciences say that "there is no such thing as race" they mean, correctly, that there are no discrete categories that come to a discrete beginning or end, that there is nothing mutually exclusive about our current (or past) categories of "race," and that there is more genetic variation within categories of "race" than between. All this is true. However, when Scotland Yard, or the Birmingham police force in England, or the New York City police force want to narrow the list of suspects in a crime, they are not primarily concerned with tight taxonomic systems of classification with no overlapping categories. That is the stuff of theoretical physics and logic in philosophy, not the practical stuff of helping to solve a crime or the practical application of molecular genetics to health delivery via genetic screening, and all the messy overlapping categories that will inevitably be involved with such enterprises. That is, some African Americans have Cystic Fibrosis even though the likelihood is far greater among Americans of North European descent, and in a parallel if not symmetrical way some American whites have Sickle Cell Anemia even though the likelihood is far greater among Americans of West African descent.

But in the world of cost-effective decision making, genetic screening for these disorders is routinely done based on commonsense versions of the phenotype. The same is true for the quite practical matter of naming suspects.

SEARCHING FOR RACIAL AND ETHNIC MARKERS IN FORENSIC DNA

In the July 8, 1995, issue of the *New Scientist*, entitled "Genes in Black and White," some extraordinary claims were made about what is possible to learn about socially defined categories of race from reviewing information gathered using new molecular genetic technology. In 1993, a British forensic scientist published what is perhaps the first DNA test explicitly acknowledged to provide "intelligence information" along "ethnic" lines for "investigators of unsolved crimes." Ian Evett, of the Home Office's forensic science laboratory in Birmingham, and his colleagues in the Metropolitan Police, claimed that their DNA test can distinguish between "Caucasians" and "Afro-Caribbeans" in nearly 85 percent of cases. Evett's work, published in the *Journal of Forensic Science Society*, draws on apparent genetic differences in three sections of human DNA.[12] Like most stretches of human DNA used for forensic typing, each of these three regions differs widely from person to person, irrespective of race. But by looking at all three, the researchers claimed that under select circumstances it is possible to estimate the probability that someone belongs to a particular racial group. The implications of this for determining, for practical purposes, who is and who is not "officially" a member of some racial or ethnic category are profound. The legal and social uses of these technologies are already in considerable use by the *cognoscenti*, and they are poised to "take off."

Here is an example. More than a decade ago, several states began keeping DNA database files for sexual offenders. Three factors converged to make this a popular decision by criminal justice officials that would be backed by politicians and the public: (1) sex offenders are those most likely to leave body tissue and fluids at the crime scene; (2) they rank among the most likely repeat offenders; and (3) their crimes are often particularly reprehensible in that they violate persons, from rape to molestation, and abuse the young and most vulnerable. Today, all fifty states store

DNA samples of sex offenders, and most states do the same for convicted murderers. But now thirty-four states store DNA samples of all felons.[13]

While thirty-nine states permit expungement of samples if charges are dropped, almost all of those states place the burden on the individual to initiate expungement. Thus, civil privacy protection, which in the default mode would place the burden on the state, is reversed. In other words, instead of "innocent until proven guilty" it has become "criminally suspect until proven innocent," so to speak. Twenty states now authorize the use of databanks for research to develop new forensic techniques. With the statutory language in several of those states, this could easily mean assaying genes or loci that contain predictive information—even though current usage is supposed to be restricted to analyzing portions of the DNA that are only useful as identifying markers. Since most states retain the full DNA (and every cell contains all the DNA information), it is a small step to using these DNA banks for other purposes. The original purpose has long been pushed to the background, and the "creep" expands not only to other crimes besides sexual offenses, but to misdemeanors and even those merely arrested as well.

CALIFORNIA AS A CASE IN POINT

On January 5, 2006, US President George Bush signed into law HR 3402, the Department of Justice Reauthorization bill of the Violence Against Women Act (VAWA) of 2005. This legislation for the first time permits state and federal law enforcement officials the right to transfer DNA profiles of those merely arrested for federal crimes into the federal Combined DNA Index System (CODIS) database. Previously, only convicted felons could be included. Those DNA profiles will remain in the database unless and until those who are exonerated or never charged with the crime request that their DNA be expunged. Thus the default will be to store these profiles, and expunging requires the proactive agency and financial resources of those arrested.

This announcement was the source of celebration by one of the leading providers of DNA testing services, Orchid Cellmak, Inc., of Princeton, New Jersey. The President and Chief Executive Officer of Cellmak, Paul J. Kelly, immediately issued a statement applauding this development:

> This is landmark legislation that we believe has the potential to greatly expand the utility of DNA testing to help prevent as well as solve crime. . . . It has been shown that many perpetrators of minor offenses graduate to more violent crimes, and we believe that this new legislation is a critical step in further harnessing the power of DNA to apprehend criminals much sooner and far more effectively than is possible today.[14]

But there is yet another reason why we must be much more wary of these developments. Criminologists and statisticians have provided enough convincing evidence that reliability may be a systemic issue with regard to "exact matches," leading to false "hits" with traditional STR approaches.[15] As for the possibility of using full DNA samples for forensic research, attempts to determine physical features, such as skin color, hair texture, and eye pigment, have already been made.[16] These techniques, because they rely on "admixture estimates" discussed earlier, are also rife with reliability issues despite their veneer of exact precision with regard to continental genetic affinity, or, put bluntly, racial diagnosis. This kind of categorizing of subjects and patients is occurring in medical and health journals, often with the idea that pharmaceuticals could be tailored to patients according to putative notions of their ancestral genetic "admixture." Researchers are also finding new ways to identify genetic variants related to "admixed" populations that they believe may be "linked" to variable complex disease conditions, such as end-stage renal disease.[17] Here, whole areas of the genome are assumed to be ancestrally "African" or "European" with very little discussion of how such prior determinations of purity are, or are not, relevant for all self-identified Africans and Europeans.

AN UNREGULATED NO MAN'S LAND: NO OVERSIGHT, NO GUIDELINES

Much like the industry of assisted reproduction in the United States, there is a complete absence of regulation or quality control with genetic ancestry testing. There is no requirement for transparency in the construction and use of reference populations. Any company can claim that their laboratories can analyze your DNA to provide accurate information about your ancestry. If three different companies provide three different answers (as in the *60 Minutes* report noted at the outset), what is a consumer to do? Which company is correct, or, more to the point, which one is more likely to be correct? There is no way of knowing, since we have no "gold standard" for excellence or professional self-policing. This was pointed out in *Science* four years ago,[18] and in November 2008, the American Society of Human Genetics (ASHG) issued a statement on ancestry testing that included five recommendations emphasizing the need for greater responsibility, research, explanatory clarity, collaboration, and accountability by these direct-to-consumer companies.[19] The statement also pointedly warned of several important limitations to the scientific approaches used to infer genetic ancestry, including the false assumption that contemporary groups are reliable substitutes for ancestral populations, and most significantly, the lack of transparency regarding the statistical methods that companies use to determine test results.[20]

But while the ASHG statement calls for greater transparency, we have seen that private sector providers of ancestry testing have proprietary reasons for keeping secret their own particular combinations of key technology, software, and population sampling procedures. Most are unwilling to disclose the size and composition of their reference populations. Without mechanisms to enforce transparency, there is no way of assessing the scientific basis for specific assertions of "percent ancestry." For example, until and unless there is a publicly available version of what constitutes a 10 percent European or a 100 percent African, etc., claims about 80 percent ancestry cannot be fully understood or tested, much less replicated.

Building on the ASHG recommendations for transparency, there is a need for specific policies enforced by federal agencies. For example, the Federal Trade Commission and the Centers for Disease Control and Prevention can and should play pivotal roles in setting industry standards for what constitutes responsible and accountable practices. These agencies can promote the research necessary to identify minimal guidelines for presenting the fair uses and clear limitations of current genomic technologies. Guidelines for transparency would also include clear statements spelling out the risks associated with over-extrapolating or misinterpreting genetic ancestry results. The active involvement of regulatory agencies would provide infrastructure for the interdisciplinary dialogue necessary to create effective policies and for maintaining industry standards.[21] While supporting such measures, we should not be naïve about their effectiveness, since the demands on these companies to generate profits are strong and insistent. It is difficult to exaggerate the role that money plays in this whole process, whether for ancestry testing companies trying to stay in business or members of groups seeking to cash in on casino gambling by being designated an Indian tribe by the US Interior Department's Bureau of Indian Affairs (BIA). For example, the Mashantucket Pequots (largely African Americans) and the Mohegan Tribal Nation (largely whites) of Connecticut have established the largest casinos in the world, generating billions of dollars in profits.

CONCLUSION

There are certain things that we can know by using DNA to authenticate ancestral links. The most important and the most definitive tests reveal genetic links (or lack thereof) to parents and grandparents, for example, deploying Y chromosome analysis to determine direct paternal lineage and mtDNA to determine the maternal line. However, claims to determine links to ancestral populations of many prior centuries must be necessarily incomplete, tentative, speculative, and of limited use. The choice of markers is completely contingent upon the choice of the reference population and profoundly limited by sampling, where researchers must proceed with untested (and sometimes untestable) assumptions to leap across unknown time and space barriers. Since we are witnessing a surge in ancestry testing across the globe, the best advice for the unsuspecting consumer is *caveat emptor* (buyer beware).

NOTES

1. Deborah A. Bolnick, Duana Fullwiley, Troy Duster et al., "The Science and Business of Genetic Ancestry Testing," *Science* 18 (October 19, 2007): 399–400.
2. Victor B. Penchaszadeh, "Abduction of Children of Political Dissidents in Argentina and the Role of Human Genetics in Their Restitution," *Journal of Public Health Policy* 13, no. 3 (autumn 1992): 291–305.
3. Amy Harmon, "DNA Gatherers Hit a Snag: The Tribes Don't Trust Them," *New York Times*, December 10, 2006.

4. Josephine Johnston, "Resisting a Genetic Identity: The Black Seminoles and Genetic Tests of Ancestry," *Journal of Law, Medicine and Ethics* 31 (November 2003): 262–71.
5. Mark D. Shriver, Michael W. Smith, Li Jin et al., "Ethnic-affiliation Estimation by Use of Population-specific DNA Markers," *American Journal of Human Genetics* 60 (1997): 957–64.
6. Duana Fullwiley, "The Biologistical Construction of Race: Admixture Technology and the New Genetic Medicine," *Social Studies of Science* 38, no. 5 (2008): 695–735.
7. Duana Fullwiley, "The Molecularization of Race: Institutionalizing Human Difference in Pharmacogenetics Practice," *Science as Culture* 16, no. 1 (March 2007): 1–30; Troy Duster, "The Molecular Reinscription of Race: Unanticipated Issues in Biotechnology and Forensic Science," *Patterns of Prejudice* 40, nos. 4/5 (2006).
8. See Pilar N. Ossorio, "Myth and Mystification," chapter 9 in this volume.
9. The Native American Tribal Organizations and Councils are part of the federally recognized government-to-government partnership with the US General Services Administration.
10. Kenneth M. Weiss and Jeffrey C. Long, "Non-Darwinian Estimation: My Ancestors, My Gene's Ancestors," *Genome Research* 19 (May 4, 2009): 703–10.
11. I. W. Evett, I. S. Buckleton, A. Raymond et al., "The Evidential Value of DNA Profiles," *Journal of the Forensic Science Society* 33, no. 4 (1993): 243–44; I. W. Evett, "Criminalistics: The Future of Expertise," *Journal of the Forensic Science Society* 33, no. 3 (1993): 173–78; I. W. Evett, P. D. Gill, J. K. Scranage et al., "Establishing the Robustness of Short-Tandem-Repeat Statistics for Forensic Application," *American Journal of Human Genetics* 58 (1996): 398–407; Alex L. Lowe, Andrew Urquhart, Lindsey A. Foreman et al., "Inferring Ethnic Origin by Means of an STR Profile," *Forensic Science International* 119 (2001):17–22.
12. I. W. Evett, "Criminalistics."
13. Tania Simoncelli, "Dangerous Excursions: The Case Against Expanding Forensic DNA Databases to Innocent Persons," *Journal of Law Medicine & Ethics* 34 (summer 2006): 390–97.
14. Orchid Press Release, January 6, 2006, 1.
15. William C. Thompson, "The Potential for Error in Forensic DNA Testing and How that Complicates the Use of DNA Databases for Criminal Identification," Council for Responsible Genetics national conference, "Forensic DNA Databases and Race: Issues, Abuses and Actions," June 19–20, 2008, New York University, www.gene-watch.org.
16. Duana Fullwiley, "Can DNA Witness Race: Forensic Uses of an Imperfect Ancestry Technology," chapter 6 in this volume.
17. W. H. Linda Kao, Michael J. Klag, Lucy A. Meoni et al., "*MYH9* is Associated with Nondiabetic End-stage Renal Disease in African Americans," *Nature Genetics* 40 (October 2008): 1185–92.
18. Bolnick, "The Science and Business."

19. The American Society of Human Genetics Ancestry Testing Statement, November 13, 2008, www.ashg.org.
20. Sandra Soo-Jin Lee, Deborah Bolnick, Troy Duster, et al. "The Illusive Gold Standard in Genetic Ancestry Testing." *Science* 325 (July 3, 2009): 38–39.
21. Ibid.

6

CAN DNA "WITNESS" RACE?

FORENSIC USES OF AN IMPERFECT ANCESTRY TESTING TECHNOLOGY

Duana Fullwiley

In courts of law an intelligent evaluation of facts is often difficult or impossible without the application of some scientific, technical, or other specialized knowledge. The most common source of this knowledge is the expert witness.[1] On August 11, 2004, an African American man named Derrick Todd Lee was convicted for the first of a series of murder and rape cases in South Louisiana. The life sentence he received would be followed by a death penalty ruling in a second case, which included evidence from several others, just a few months later. Lee had a record of questionable peeping behavior, domestic violence, burglary, and assault on his "rap sheet" dating back to his youth. According to various accounts, despite several encounters with the law, Lee fell through the cracks as the local police attempted to compile a profile of the perpetrator of the violent serial deaths of seven women in the Baton Rouge area in the early 2000s.[2]

Lee's eventual convictions were largely based on his Y chromosome STR DNA profile that matched DNA from samples found on mulitple victims' bodies. The rulings were also based on a riveting testimony and identification of Lee by a would-be victim who narrowly escaped a rape attempt at his hands. One other identification "match" in the first of Lee's two trials was based on bloody footprints left on flooring that matched his construction boots at the scene of one crime, and the assessment of those boots as Lee's by his former girlfriend and his fifteen-year-old son.[3]

Most of the evidence compiled and marshaled by the prosecution easily resembles standard forensic details that appear in criminal cases of violent aggression. What is less typical about the way that the Louisiana police and FBI task force narrowed the focus on Derrick Todd Lee is the series of events preceding the procurement of his DNA through a cheek swab. Lee was the first person in the United States to be identified as a possible suspect by an unconventional DNA analysis tool that racially profiled the DNA he left at a crime scene.

The technology that purported to read Lee's race in his DNA is trademarked as DNAWitness. The name is not accidental. Its inventors at the now-defunct DNAPrint Genomics, Inc., wanted to convey the idea that this technology itself could embody the power of the "expert witness," thereby possessing the ability to "call out" the perpetrator, through literal genotyping "call outs" of his specific DNA base pairs. Forensic analyses with DNAWitness are based on a comparison of a sample of unknown origin with a panel of genetic markers called Ancestry Informative Markers, or AIMs. The basic process of an AIMs analysis consists of a comparative exhibition of varying autosomal coding markers and their relative frequencies in four world populations. The goal of this specific iteration of the AIMs test, packaged only for forensics as DNAWitness, is to infer the aggregate of phenotypes associated with any one racial category in the United States. Such an inference is based on the extent to which the anonymous sample expresses allelic variations of markers comprising a panel that is thought to differ in people from the continents of Africa, Asia, Europe, and (pre-Columbian) America, who are called "parental populations." Geneticists Kenneth Weiss and Jeffrey Long provide a detailed critique of the bias and scientific oversights of admixture technologies such as this one. They write:

> Whether the investigator uses external information or makes estimates from the samples at hand, the parental populations are abstractions that conform to only the simplest kind of genetic structure. This structure places heavy emphasis on the idea that the world once harbored distinct and independently evolved populations that have now undergone admixture of an unstated type (often seeming to connote admixture due to colonial era migrations). Regardless of the intent, this idea of population structure is unfortunately more in line with race concepts held by

European explorers and traders than with the recent genetic evidence supporting the serial sampling of human evolutionary history.[4]

In the case of the south Louisiana serial killer, DNAWitness yielded "ancestry estimates" that the perpetrator's genetic make up was 85 percent sub-Saharan African and 15 percent Native American. The Louisiana task force's previous search for a "Caucasian" male was thereafter deemed to be potentially off the mark. The suspect, as deduced by DNAWitness, was most likely a "lighter skinned black man" which was inferred, via a genetic structure analysis, from probabilistic ancestry percentages revealed in the perpetrator's DNA.

In this chapter, I examine the use of DNAWitness to determine the prospective race of a suspect in order to provide evidence to law enforcement for narrowing a suspect pool. I argue that DNAWitness falls short of legal and scientific standards for trial admissibility, while it eludes certain legal logics with regard to the use of racial categories in interpreting DNA. DNAWitness can offer vague profiles in many cases and has a wide margin of error that also often absorbs what might be understood to be important aspects (i.e., substantial percentages) of ancestral heritage or of a forensic "racial profile." Moreover, this technology's individual ancestry estimates are highly vulnerable to social and political interpretations of phenotype, and may be impossible to accurately interpret with a sufficient degree of objectivity, which is required of both science and law. It is possible, however, that this test may help to predict a range of skin color phenotypes, as was the case for Lee, since many of the AIMs are skin and hair pigmentation alleles.

The AIMs technology (packaged with different names depending on the market and client) as first manufactured for commercial forensics by DNAPrint Genomics is specifically designed to assess allelic frequency differences of coding DNA, or Single Nucleotide Polymorphisms (SNPs). This is important since markers that the test makers interpret as "African," or "European," for example, are also found in other world populations that differ from the prior continental referent populations used by the company (African, European, Native American, and Asian) in name and geographic location. This is to say that differences in ancestry profiles may be due to selection, gene flow, genetic convergence, or genetic drift. The presentation of DNAWitness test results demonstrates no attempt to distinguish between these different mechanisms of locus possession in individuals or in groups. Direct and unique ancestry (gene flow) is but one among several mechanisms that might explain shared sequence variation among and between racialized individuals. The simple description of a certain frequency, or set of frequencies, as "African" ancestry may constitute a false designation of "racial type," while, conversely, it might not. The fact that there is no gold standard for these types of technologies (which now include techniques called "molecular photofitting" (see note 18) should make the legal community pause before lauding their effects.

SCIENTIFIC MERIT IN SCIENCE AND LAW

From the outset, before evaluating the scientific criteria for admissibility in a trial setting, it must be clarified that DNAWitness has not been used at the trial stage, but rather at the pretrial stage as *prospective* information for investigating officers. Nonetheless, it is critical to consider the scientific standards for legal admissibility to shed light on the ways in which this technology may actually do harm in the courtroom, since its scientific shortcomings can be easily identified with regard to admissibility rules. Furthermore, holding this technology to accepted legal standards with regard to "expert" use of science and technology will also allow us to better understand DNAWitness's problematic role in the legal setting at *any* stage.

Legal precedent would have us focus on three federal cases to determine how standards derived from norms of scientific merit constitute the rules for admissibility in a court of law. These are *Daubert v. Merrell Dow Pharmaceuticals*, *General Electric & Co. v. Joiner*, and *Kumho Tire Co., Ltd v. Carmichael*. Issues of "reliability," "scientific validity," and whether techniques "can be tested" and "falsified" are of critical concern. As stated in *Daubert v. Merrell Dow*, "scientific methodology today is based on generating hypotheses and testing them to see if they can be falsified; indeed, this methodology is what distinguishes science from other fields of human inquiry."[5] More specifically, a "non-exclusive checklist for trial courts to use in assessing the reliability of scientific expert testimony," provided in Notes to 702, Federal Rules of Evidence, include: (1) whether the expert's technique or theory can be challenged in some objective sense, or whether it is instead simply a subjective, conclusory approach that cannot be reasonably assessed for reliability; (2) whether the technique or theory has been subject to peer review and publication; (3) whether the rate of error of the technique or theory when applied is known; (4) whether there exists maintenance of standards and controls; and (5) whether the technique or theory has been generally accepted in the scientific community."[6] As will become clear below, DNAWitness fails to meet this basic checklist on several counts. These standards were established for the use of scientific evidence in a court of law independent of DNA testing, yet they nonetheless hold for all scientific evidence.[7] Effective December 1, 2000, several amendments to the federal rules, namely, with regard to procedure and methods of reliability, made it clear to both the bench and bar "that an attack on the procedure used to test DNA for evidentiary purposes can be an effective challenge to the weight of any DNA evidence admitted."[8] Thus, presenting genetic results in less than exact and recognized ways could prove detrimental to case arguments.

HISTORICAL BACKGROUND: "RACE" MARKERS THEN AND NOW

Several of the highly "informative" markers at work in DNAWitness have long been the focus of human variation studies in physical anthropology and physiology dating back to mid-twentieth-century US medicine. Biogenetic studies starting in the 1950s centered on the genetics of blood protein phenotypes that differed between Americans of "African" descent and those of "European stock." During this same time period, vocal social scientists of the day issued statements warning against using such traits to emphasize the (superficial) nature of phenotypes, given the breadth of the human species' biological commonality. The most noteworthy of such public discussions on the issue was the 1950 UNESCO Statement on Race. Written largely by social scientists, yet reviewed by leading geneticists, the authors centered on the "temporary" nature of "varying manifestations" of traits. Their goal was to unsettle the common belief of the day in racial biological determinism and fixity:

> In looking at the different "races" of mankind today . . . the varying manifestations of physical traits which they exhibit are not "end results"

> but bills of exchange, as it were, drawn on the bank of time, negotiable securities which can be turned into the coin of any realm with which it is sought to have biological relations. In other words, we perceive the consequences of different histories of biological experiences in the "races" of today.[9]

Still many researchers continued to examine such "end results"—as ends in and of themselves—that "varyingly manifested" in blacks and whites. Nonetheless, even in the 1960s, some scientists working with such physical traits (which today are characterized as some of the "most informative" of the Ancestry Informative Markers operative in DNAWitness) provided clear caveats against assumptions that the frequent possession of certain alleles in a referent group and in a comparative sample necessarily indicated direct and unique ancestry *from* the referent group *to* the sample. In fact, to draw such a conclusion of gene flow, one researcher in the 1960s wrote that a number of facts needed to be held constant to infer ancestry—facts regarding the certainty of a population's putatively fixed characteristics, both genetic and ethnic. Zoologist, anthropologist, and pediatrician T. Edward Reed, who provided a detailed review of eleven studies on such markers that had taken place through 1969, warned that researchers had not "always appreciated" the simple and obvious criteria that must be met to estimate ancestry from what are now called Ancestry Informative Markers. He wrote:

> Critical evaluation of estimates of M [Caucasian mixture in "American Negroes"] requires complete specification of the needed criteria and judgment on the degree to which these criteria are met. These criteria are simple and obvious, but the demands they make have not always been appreciated. They are as follows:
>
> 1) The exact ethnic compositions of the two ancestral populations, African Negro and Caucasian, are known;
>
> 2) No change in gene frequency (for the gene in question) between ancestral and modern populations either of African Negroes or of American Caucasians has occurred;
>
> 3) Interbreeding of the two ancestral populations is *the only factor affecting gene frequency* in U.S. Negroes—that is, there has been no selection, mutation, or genetic drift.[10]

As stated in the introductory section, "admixture" analyses (based on the use of several of these same markers reviewed by Reed and later used in DNAPrint's products) specifically did not take into account these other possible sources of varied gene frequency distributions in present-day populations, which were enumerated over forty years ago. Other researchers in the twenty-first century have repeated Reed's call.[11] Yet, as concerns DNAPrint's various AIMs products, these points went overlooked. Instead, the company pursued this technology by creating a methodology that was seemingly robust if allowed to operate independent of any recognition of this larger set of criteria.

When compiling previous studies of unevenly distributed gene frequencies of a handful of known traits in "Africans," "East Asians," "Europeans," and "Native Americans," DNAPrint scientists initially identified the continental groups of interest, and then examined human DNA samples from public genetic databases, such as dbSNP, to identify thirty-four genetic markers that displayed at least a 50 percent allelic frequency distribution difference between any two of the four groups. To determine this, all subjects in a defined group, for example "Africans," were tested for the panel of alleles. Then, the researchers determined the frequency at which AIMs appeared in each group as a whole. These referent populations surely displayed as much genetic heterogeneity as any group, but the point of the AIMs research was to find those specific alleles that they shared at a certain frequency, as long as that frequency, when contrasted with at least one other of the four groups, demonstrated a difference of 50 percent. Such purified, or artificially homogenized, samples came to serve as "parental populations" of Africans, Europeans, Native Americans, and East Asians for "contemporary" test takers.

One image on the DNAWitness portion of the DNAPrint website illustrated this point, while keeping both "admixture" and the concepts of population purity it depends on simultaneously in play. It featured a grey world terrain map of four landmasses: (1) North and South America, (2) Europe, (3) Asia, and (4) Africa. One singular human face was superimposed on each geographical space, four faces in total. These were assumed to be American test takers, who, even in the twenty-first century, could be more or less racially allocated to the continents to which they were pegged. Though the images appeared as isolated (lone figures separated by oceans), each face is accompanied by admixture readouts and ethnic composition statistics. All were flanked by a serial code overhead and a bar graph of continental ancestry percentages below. A brown female face was placed squarely between North and South America and is the only one made to represent two continents on the map. Her admixture readout revealed that she is 50 percent "European," 45 percent "Native American," and 5 percent "African." We, the viewers, were not told if she self-identified as "Latino" or "Native American," and as such she was made to do the work of being both. An "African American" female image and an "Asian American" male face were similarly accompanied by readouts of their admixtures of the continental ancestral geographies and populations over which they are superimposed (Africa and Asia respectively), with small percentages of other populations' genes that had "flowed" into them ("European" in both cases). The "European American" face was the only one that is revealed to be "100 percent," in this case, of course, pure "European."

AIMs analyses most often rely on a series of unknowns, which are the three points that Reed highlighted as necessary to proceeding with confidence with this model. When researchers cannot answer questions, such as "the exact ethnic composition" of the contributing parental source populations or even how many sources there were, in any certain terms, they instead rely on assumptions. These assumptions are not arbitrary. They are usually informed by a given society's historical account of when the two or more groups, who are the putative "parentals," encountered each other. But problems arise when social context and historical accounts meet with sampling limits. For instance, using contemporary, acknowledged Native American groups as one source population for Mexican Americans might overlook an actual history of genocide of peoples who no doubt contributed to present-day populations in the Americas, but who may no longer exist.[12] To complicate matters more, Mexican Americans often have more Amerindian heritage than the referent groups posited as their Native American ancestors. For political and historical reasons, individuals need only possess one-eighth (12.5 percent) demonstrable Native American ancestry to be considered Native American, whereas Mexicans may have considerably more.[13] Finally, in addition to these quandaries, when alleles that have a high frequency in the specific reference groups tested (those labeled "African," "European," "Native American," etc.) appear in a "client" taking the test, the AIMs test reads that the client has inherited the specific referent ancestry rather than, say, ancestry (or SNPs) from other still unsampled parts of the globe.

We may now revisit the Federal Rules of Evidence "checklist" for scientific admissibility in the legal sphere. It should be clear that the basis on which DNAPrint's "parental populations" are constructed and artificially homogenized involves a certain cultural bias, and thus subjective choice, of racial typing from the outset. This is a critique that has been levied against DNAPrint recently, and which highlights the systematic bias of the results its technologies have yielded.[14] Although various studies that have used these markers have undergone peer review, the scientific journals and the professional peers of DNAPrint scientists and their collaborators have not requested that the entire set of markers, the complete details of sampling, and the assumptions that populations' trait possessions are due to gene flow be evaluated. Thus, the full scope of this technology remains proprietary and evades peer review. The technology continues to provoke controversy due to its modeling of four archaic racial types and its claim to be able to estimate individual ancestry percentages from these types without any discussion of other mechanisms for allelic similarity.[15] Additionally, although various AIMs products' margins of error may be known, the company website lists different, smaller figures (ranging from .5 to 15 percent) than DNAPrint's sole scientific advisor has admitted to publicly (as high as 30 percent).[16] On several counts, DNAWitness fails to meet the five items of the "checklist." We must therefore question its admission in courts of law, as well as its use in pretrial assessments, if the goal is to limit societal and cultural biases from entering into judgments. In this case we must be mindful of the systematic cultural assumptions that run through admixture models, where human genetic diversity is framed as neatly parceling into American political categories of race.

CONCLUSION

The rise of new genetic technologies in the past two decades has yielded a range of scientific possibilities for the courts. Not all genetic tests perform the same kinds of tasks, and none were instituted without prolonged discussion, debate, and research consensus with regard to their reliability and consistency among scientists and law enforcement.[17] As this analysis makes clear, DNAWitness is based on Ancestry Informative Marker technology, or coding SNPs, that are largely shared among individuals and groups for varying reasons—reasons that are neither described nor acknowledged explicitly in the test results offered by DNAPrint. AIMs-based technologies like DNAWitness are attempts to model human history from a specifically American perspective in order to infer present-day humans' continental origins.[18] Such inferences are based on the extent to which any subject or sample shares a panel of alleles (or variants of alleles) that code for genomic function, such as malaria resistance, UV protection, lactase digestion, skin pigmentation, and so on. There is a range of such traits that are conserved in and shared between different peoples and populations around the globe for evolutionary, adaptive, migratory, and cultural reasons. To assume that people who share, or rather *co-possess*, these traits can necessarily be "diagnosed" with a specific source ancestry is misleading. Not only will siblings often share the same profile, but individuals from all four "parental" continental groups offered up by the model could feasibly share similar profiles, though it could just as often happen that siblings or individuals hailing from specific continental groups do not share the same profiles. As a forensics-market version of AIMs technology, DNAWitness and future developments in a similar vein, such as molecular photofitting, may offer precise mathematical ancestry percentages, but the accuracy of that precision remains debatable.

At best, this technology is an experimental modeling tool that hopes to mimic recent American human history as it reconstructs four racial types through an artificial homogenizing of markers found with relatively higher frequencies on some continents and lower frequencies on others. As compelling as DNAWitness as a tool may seem, investigators should require that DNA analyses used in the serious proceedings of law be falsifiable, reliable, and thoroughly vetted. Anything less would prove irresponsible if incorporated into criminal investigations.

NOTES

1. Notes to Fed. R. Evid. 702, www.law.cornell.edu/rules/fre/ACRule702.htm.
2. P. Roberts, *Sunday Advocate*, June 1, 2003.

3. K. O'Brien, *Times-Picayune*, August 10, 2004.
4. K. M Weiss and J. C Long, "Non-Darwinian Estimation: My Ancestors, My Genes' Ancestors," *Genome Research* 19 (2009): 703–710, at 705. The technology in question, which was once produced by DNAPrint Genomics, is also packaged as Ancestry by DNA for recreational genealogical ancestry testing. Variations of it are also used in biomedical research settings for purposes of admixture mapping for disease traits and to prevent confounding in "mixed" populations in case-control studies for complex disease traits. See www.dnaprint.com/welcome/productsandservices/index2.php (accessed March 28, 2008). Since DNAPrint ceased operations in 2009 DNA Diagnostic Center has began marketing *AncestrybyDNA*. See www.ancestrybydna.com (accessed April 19, 2010).
5. *Daubert v. Merrell Dow Pharmaceuticals*, 509 US 579 (1993).
6. Notes to Fed. R. Evid. 702.
7. E. Imwinkleried, "The Relative Priority that Should Be Assigned to Trial Stage DNA Issues," in *DNA and the Criminal Justice System: The Technology of Justice*, ed. D. Lazer (Cambridge, MA: MIT Press, 2004), 99.
8. Ibid.
9. A. Montagu, *Statement on Race* (New York: Oxford University Press, 1972), 47.
10. T. Edward Reed, *Science* 165 (1969): 762–68 (emphasis mine).
11. C. L. Pfaff, *Genetic Epidemiology* 26 (2004): 306.
12. Ibid.
13. R. Kittles and K. Weiss, *Annual Review of Genomics and Human Genetics* 4 (2003): 48.
14. Deborah A. Bolnick, Duana Fullwiley, Troy Duster et al., "The Science and Business of Genetic Ancestry Testing," *Science* 18 (October 19, 2007): 399–400; Deborah A. Bolnick, Duana Fullwiley, Jonathan Marks et al., "The Legitimacy of Genetic Ancestry Tests," *Science* 319 (February 22, 2008): 1039–40.
15. Bolnick et al., "The Science and Business"; Bolnick et al., "The Legitimacy of Genetic Ancestry Tests"; M. K. Cho and P. Sankar, reply to "Getting the Science and the Ethics Right in Forensic Genetics," *Nature Genetics* 37 (2005): 450–51.
16. For the DNAPrint Genomics account of "Accuracy and Precision," see www.ancestrybydna.com; accessed December 27, 2006. M. Shriver, "Applied population genomics: exploring the genetic architecture of recently evolved traits," presentation at Harvard University, Cambridge, MA, February 27, 2008. Shriver is the only "scientific advisor" currently listed for the company. See DNAPrint, Scientific Advisors, www.dnaprint.com/welcome/corporate/scienctificadvisors/ accessed December 27, 2006.
17. National Research Council, Committee on DNA Forensic Science, *The Evaluation of Forensic DNA Evidence, An Update* (Washington, DC: National Academy Press, 1996).
18. T. Frudakis, *Molecular Photofitting: Predicting Ancestry and Phenotype Using DNA* (Burlington, MA: Academic Press, 2007), 429.

PART IV

RACIALIZED MEDICINE

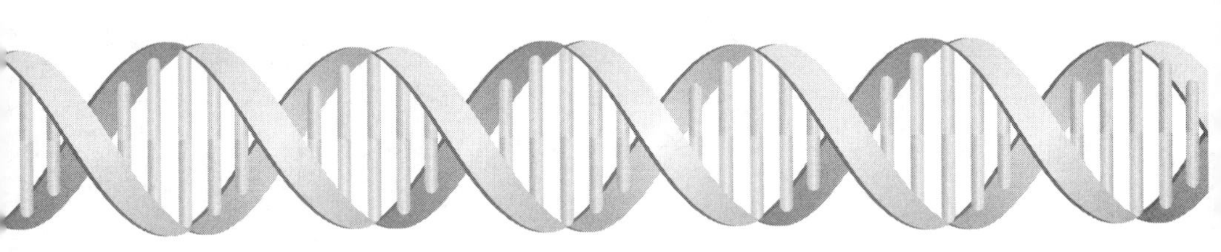

7

BIDIL AND RACIALIZED MEDICINE

Jonathan Kahn

On June 23, 2005, the Food and Drug Administration (FDA) approved a drug to treat heart failure in African Americans and *only* African Americans. This race-specific drug is called BiDil but is not a new drug. It is merely a combination into a single pill of two existing generic drugs that have been used to treat heart failure regardless of race for over a decade. BiDil was brought to the FDA by NitroMed, a hitherto small Massachusetts biotech company with no other products on the market. NitroMed explicitly requested race-specific FDA approval for its drug based on clinical data produced by its "African-American Heart Failure Trial (A-HeFT)" on the grounds that the trial population happened to be all self-identified African American.

BiDil does indeed appear to significantly help many people suffering from heart failure—a debilitating and ultimately fatal disease afflicting several million Americans. There is no scientific evidence, however, that race has anything to do with how BiDil works. This is for the simple reason that A-HeFT enrolled *only* "self identified" African Americans. With no comparison population, no legitimate claims can be sustained that BiDil works differently or better in African Americans than in anyone else. The FDA, however, accepted NitroMed's argument that because the trial population was African American then the drug should be labeled as indicated only for African Americans. This sends the troubling and unsubstantiated message that the subject population's race was somehow a relevant biological variable in assessing the safety and efficacy of BiDil. Ominously, it also gives the federal government's imprimatur to the use of race as, in effect, a genetic or biological category. By seeking and granting approval of BiDil as a drug solely to treat African Americans, NitroMed and the FDA thus opened a Pandora's box of racial politics without fully appreciating the implications of what they were doing.

BASIC CONCERNS

First, most drugs on the market today were tested almost exclusively in overwhelmingly white male populations. But we do not call these "white" drugs—nor should we. Rather, the operating assumption for approving these drugs was that the unmarked racial category of "white" was coextensive with the category "human being." That is, a drug tested on white people was good enough for everybody. In approving BiDil as a drug only for African Americans, the FDA has also implicitly adopted an assumption that drugs tested in black people are only good for black people. This sends the unintended but nonetheless powerful message that black people are somehow less fully representative of humanity than are white people.

Second, given that the BiDil researchers admit that their drug will work in non–African Americans, the most plausible reason for conducting a race-specific clinical trial is that NitroMed holds the rights to a race-specific patent that will give them control over profits from BiDil until 2020. Of course, this hardly constitutes a sound scientific basis for designing a clinical trial, but it *is* a good economic one. An older patent, which does not refer to race, expired in 2007; if BiDil had been approved for treatment regardless of race, NitroMed's patent protection would have expired in a mere two years.[1]

Third, there is the problem of who "counts" as African American. In an increasingly intermixed and complex society, one might ask just "how much" of an African American one has to be to get the drug—one-half, one-quarter (quadroon), one-eighth (octoroon)? This starts sounding suspiciously like the blood quantum thinking of the Jim Crow era. Moreover, the label itself also refers to "black" people. Does this include dark-skinned peoples from South Asia, or Australian Aborigines? The trials and label of BiDil are based on the concept of "self-identification." Self-identification, however, is a subjective social judgment. It has nothing to do with the biological phenomena of drug metabolism and response.

Fourth, race-specific labeling will make it more likely that non–African Americans who might benefit from the drug will not get it. Health care providers simply may not think of prescribing it to non–African Americans, and insurance carriers may not cover such "off-label" use. Alternatively, what happens if someone who would typically be socially identified as white decides to self-identify as African American in order to get insurance coverage for the drug? Are insurance companies then going to become arbiters of racial identity?

Fifth, marketing a race-specific drug can lead to a misallocation of health care resources. To the extent that we reduce the very real racial disparities that exist in health care to a function of genetic difference, we risk diverting political will and economic support away from addressing the pressing social, economic, and political causes of racial inequality in our society. Racialized medicine presents us with the superficially appealing and misguided message that instead of fixing injustice, we can simply fix molecules.

Finally, if the FDA approves BiDil only for African Americans, it will be giving the federal government's stamp of approval to using race as, in effect, a genetic category. But race is not genetic, as even the BiDil researchers admit. Moreover, there is no accepted biological definition of race. Given our nation's troubled history of racial oppression, this is not something that should be taken lightly.

WHAT IS RACIALIZED MEDICINE?

At the outset, it is important to distinguish between the use of race in medical practice as opposed to racialized medicine. It may be entirely appropriate, even necessary, to use race when tracking and addressing broad issues such as health disparities in American society. Understanding race as a social construct is entirely consistent with recognizing and addressing race-based inequalities in access to or quality of medical care in our society. Such inequalities reflect the biological implications of the social and historical phenomenon of racial discrimination.

Social understandings of race vary over time and across space. In the past, the US Census has included racial categories ranging from Mulatto to Hindu. In the Jim Crow South, children of Armenian or Greek immigrants were sometimes mandated to go to schools designated for black children. Today, someone of light brown complexion who is socially identified as "black" in the United States (say, for example, someone with a white mother from Kansas and a black father from Kenya) might be identified as "white" in Jamaica or Brazil. In the United States, the concept of "self-identification" has become the norm in assigning racial identities to individuals. As a social practice for collecting census data this makes sense. But as a medical or scientific practice it is far more problematic. In medical practice what matters is our shifting understanding of the correlations between such evolving social identities and the evolving economic, political, and environmental conditions to which they may be related. For example, what are we to make of the fact that African Americans suffer from disproportionately high rates of hypertension, but Africans in Nigeria have among the world's lowest rates of hypertension, far lower than the overwhelmingly white population of Germany?[2] Genetics certainly plays a role in hypertension. But any role it plays in explaining such differences must surely be vanishingly small.

There may be occasions where race can be productively used even in genetic research, but in such cases it is very important to differentiate between using a racial group to characterize a gene versus using a gene to characterize a racial group. Thus, for example, a researcher trying to understand the genetics of diabetes may choose to study the Pima Indians in the Southwest United States because that group has a very high incidence of diabetes. This is an example of using a socially identified racial or ethnic group to try to characterize a gene (here, for diabetes). It is quite another thing, however, for a researcher who finds such a gene to use it to characterize the identity of Pima Indians as a group with "the gene for diabetes." The former use does not necessarily stigmatize or define a group by its genetics, the latter use does.

In such situations, medical practitioners need not, indeed often should not, ignore race. The issue is not primarily one of *whether* to use racial categories in medical practice but *how*. Carefully taking account of race to help understand broader social or environmental factors that may be influencing health disparities can be warranted in certain situations. But it is always important to understand that race itself is not an inherent causal factor in such conditions and that race is itself a socially constructed term, not a biological reality.

In contrast, *racialized* medicine is premised on an implicit, and sometimes explicit, understanding of race as a genetic construct. Such an understanding is both scientifically flawed and politically dangerous. Since the inception of the Human Genome Project, much time and attention has been devoted to ensuring that biological knowledge emerging from advances in genetic research is not used inappropriately to make socially constructed racial categories appear biologically given or natural. Since Richard Lewontin's groundbreaking work on blood group polymorphisms in different groups and races in the 1970s, scientists have understood that race will statistically explain only a small portion of genetic variations. As a 2001 editorial in the journal *Nature Genetics* put it, "Scientists have long been saying that at the genetic level there is more variation between two individuals in the same population than between populations and that there is no biological basis for 'race.'"[3] More recently, an editorial in *Nature Biotechnology* asserted, "Race is simply a poor proxy for the environmental and genetic causes of disease or drug response. . . . Pooling people in race silos is akin to zoologists grouping raccoons, tigers and okapis on the basis that they are all stripey."[4]

Politically, history teaches us that constructing races as genetically bounded and discrete categories is only one short step from constructing races as inferior and superior. Racism feeds on biologically reductive constructions of racial difference. It is imperative to recognize the significance of race to understand and address the real and persistent health disparities that plague our country. But these disparities are the result of social, economic, and political histories of injustice. They demand social, economic, and political responses. If we falsely reduce health disparities among socially defined racial groups to a function of genetic difference to be addressed through race-specific medicine, we risk diverting valuable resources away from developing policies and practices to confront the true causes of health disparities.

Unlike *racialized* medicine, which treats race as genetic, the *use* of race in medical practice has many legitimate and important places. Collecting broad-based epidemiological data is perhaps foremost among these. Only by using social categories of race is it possible to identify and track racial disparities in health, health care access, and outcomes. Such information is needed to address ongoing issues of racial injustice in society. It may also be appropriate for individual health practitioners to take race into account under certain circumstances in trying to assess the needs of their patients. To the extent that health practitioners understand that race, as a social phenomenon, has biological consequences—such as where higher incidence of hypertension might in part be due to an array of environmental, social, or economic factors disproportionately associated with being a racial minority in the United States—it may be legitimate and important to take race into account in formulating appropriate medical interventions.

BIDIL: A CAUTIONARY TALE OF EXPLOITING RACE IN DRUG DEVELOPMENT

Why then did NitroMed seek race-specific approval for its drug? At the FDA approval hearing much was made of following the "signal" from two trials conducted in the 1980s that first tested the generic components of BiDil as a treatment for heart failure. As discussions at the hearings progressed, however, it became clear that reviewers were relying primarily on data from the first trial, which placed only forty-nine African Americans on the two BiDil generic components. Given that most valid clinical trials test a drug in thousands of subjects, results from forty-nine African Americans seem a slender reed indeed upon which to weigh the value of a drug. Perhaps then the answer lies not in such tenuous medical evidence but in stouter commercial considerations. It turns out that NitroMed holds two key patents for BiDil. The first covers the non-race-specific use of BiDil. This patent, however, expired in 2007. The second patent is race-specific and just happens to last until 2020. (How and why a race-specific patent was granted will be addressed in the next section.) The extra thirteen years of patent protection may present a compelling commercial reason for seeking to cast BiDil as a racial drug, but it is not supported by the medical evidence.

BIDIL'S ORIGINS

How did we get to this point? If we go back to its origins, we find that BiDil did not begin as an ethnic drug. Rather it became ethnic over time and through a complex array of legal, commercial, and medical interventions that transformed the drug's identity. Over the past twenty years a revolution has occurred in heart failure treatment with the development of a wide array of pharmaceutical interventions to improve both the quality of life and longevity of people suffering from heart failure. One of the earliest breakthroughs came in the 1980s with the first Vasodilator Heart Failure Trial (V-HeFT I). This trial lasted from 1980 to 1985. It was led by Dr. Jay Cohn of the University of Minnesota and involved cardiologists from around the country working together with the US Veterans Administration. The trials found that patients receiving a combination of two vasodilators called hydralazine and isosorbide dinitrate (H/I) seemed to have a lower rate of mortality.[5] These generic drugs—H/I—would later become BiDil. It is this trial that placed forty-nine African Americans on the H/I combination that would later become the "signal" followed by NitroMed in seeking race-specific approval for BiDil.

The V-HeFT I trial was soon followed by V-HeFT II, which lasted from 1986 to 1989.[6] This trial compared the efficacy of the H/I combination against the drug enalapril, an ACE inhibitor. (ACE inhibitors, or angiotensin-converting enzyme inhibitors, are a group of pharmaceuticals that are used primarily in the treatment of hypertension and congestive heart failure.) It found an even more pronounced beneficial effect on mortality in the enalapril group, establishing ACE inhibitors as a front line therapy for heart failure. ACE inhibitors, however, did not totally supplant H/I because not everyone responds well to them and some others cannot tolerate the side effects.

The V-HeFT investigators did not build the trials around race or ethnicity. They enrolled both black and white patients but in the published reports of the trials' successes they did not break down the data by race. Rather, they presented H/I (the BiDil drugs) as generally efficacious in the population at large, without regard to race.

THE LEGAL AND COMMERCIAL CONSTRUCTION OF BIDIL AS A RACE-SPECIFIC DRUG

The role of law as player in the emergence of BiDil as a race-specific drug began in 1980, more or less coincidentally with the initiation of V-HeFT I. That year, President Jimmy Carter signed into law two pieces of legislation that would come to transform relations between industry and academic researchers. The first, the Stevenson-Wydler Technology Innovation Act of 1980 (15 U.S.C. § 3701), encouraged interaction and cooperation among government laboratories, universities, big industries, and small businesses. The second, the Patent and Trademark Laws Amendment Act of 1980 (35 U.S.C. §§ 200–212), commonly known as the "Bayh-Dole Act, allowed institutions conducting research with federal funds, such as universities, to retain the intellectual property rights to their discoveries. It is in this context that the research findings of V-HeFT, produced in cooperation with the United States Veterans Administration, could be commercialized through patent and trademark law. Thus, lead cardiologists in the V-Heft trials, Jay Cohn and Peter Carson, later were able to obtain intellectual property rights in BiDil-related patents and thereupon enter into deals with the likes of NitroMed to commercialize the discoveries made through the V-HeFT trials.

In 1989, Cohn obtained a patent on a "method of reducing mortality associated with congestive heart failure using hydralazine and isosorbide dinitrate," (U.S. Pat. 4,868,179). He then licensed the rights to a company called Medco, which developed BiDil as a *new* drug, a combination of H/I in single dose form. BiDil was a breakthrough of convenience—it made it easier to use and dispense the drug—but it was not itself a new therapy. Again, at this point it was still a drug for everyone, *regardless of race*. In the early 1990s Medco invested time and money to conduct bioequivalence tests and develop marketing strategies in preparation for submitting a New Drug Application (NDA) for BiDil to the FDA in 1996.

Even at this early stage the true breakthrough of BiDil was not the combination of two generic drugs into a single pill, it was the development of new intellectual property rights whose value was contingent on FDA approval of the new drug. With a patentable therapy in hand, drug companies would have an incentive to educate physicians and market the new drug. In contrast to the classic justification for patents as incentives to develop new drugs, intellectual property rights here provided instead an incentive for developing a new marketing strategy based on existing therapy. Moreover, given that the two drugs comprising BiDil were already available as generics, this also indicates how patent law and regulatory approval may distort a market, potentially obscuring less expensive generic alternatives that have the same therapeutic value.

The FDA ultimately rejected this first NDA in 1997 because it found the retrospective analysis of data from the V-Heft trial was insufficiently powered to meet the regulatory criteria of statistical significance. It is important to note here that the FDA advisory committee reviewing the drug did not think BiDil didn't work. To the contrary, many of the doctors on the panel were generally convinced of its clinical efficacy. They turned down the application because V-HeFT trials were not designed as *new* drug trials and so the data they produced could not meet the regulatory criteria of statistical significance required for new drug approval.[7]

Following the FDA rejection in 1997, the value of the intellectual property rights to BiDil plummeted along with Medco's stock. The rights reverted to Cohn, and Medco exited the story of BiDil's development. It was at this point that Cohn, together with Carson and others, went back to the V-Heft data and broke it out by race for the first time.

The intervention of the federal regulatory system to deny the NDA marks the turning point on BiDil's journey toward ethnicity. The regulatory action taken by the Advisory Committee impelled the BiDil researchers to reconceptualize their drug along racial lines in order to get a second bite of the apple of FDA approval.[8] After the publication of an article purporting to show that BiDil worked better in African Americans,[9] the value of the intellectual property rights to BiDil rebounded—not because of any changes to the underlying molecular structure or biological effects of BiDil as a drug, but through the reanalysis of the old V-HeFT data along racial lines.

NitroMed acquired the intellectual property rights to BiDil in September 1999. In the hands of its new corporate handlers, together with their public relations consultants, BiDil soon was reborn as an ethnic drug. The subsequent spate of publicity attending the inauguration of A-HeFT (African American Heart Failure Trial) marks how the renewed value of the patent to BiDil provided an incentive for NitroMed to educate doctors and the public about the nature and value of this "new" drug for African Americans.

In the next logical extension of patent rights into the process of creating an ethnic drug, Cohn and Carson jointly filed for a new BiDil-related patent on September 8, 2000. With the title "Methods of treating and preventing congestive heart failure with hydralazine compounds and isosorbide dinitrate or isosorbide mononitrate," the patent appears much the same as Cohn's original 1989 patent. Upon closer inspection, however, the abstract to the patent specifies that the "present invention provides methods for treating and preventing mortality associated with heart failure in an *African American* patient."[10] The issuance of the new patent is commercially important because the original patent was set to expire in 2007. The new race-based patent will not expire until 2020. In this chapter of BiDil's development, *patent law did not spur the invention of a new drug, but rather the reinvention of an existing therapy as race-specific.*

In 2001, NitroMed approached the FDA with its proposal to obtain race-specific approval for BiDil. The FDA responded with a letter stating that BiDil might be "approvable" pending the successful completion of a race-specific confirmatory drug trial. With the FDA letter in hand, NitroMed was able to raise over $30 million in venture capital during the nadir of the dot-com bust in 2001 to initiate A-HeFT. A-HeFT enrolled only "self-identified African American" subjects. After a slow start, 1,050 subjects were ultimately enrolled. The trial was halted early, in July 2004, by NitroMed's Data Safety Monitoring Board because it found a striking degree of efficacy in early results indicating that BiDil reduced mortality by some 43 percent. Such strong positive results suggested that all trial participants should get the drug. NitroMed's stock price more than tripled on the announcement. This was followed in June 2005 by the FDA's approval of the new race-specific NDA for BiDil.

THE COMPLEX RACIAL POLITICS OF RACE-SPECIFIC MEDICINE

Given such striking results, it is perhaps not surprising that many prominent health activists and organizations in the African American community strongly supported the approval of BiDil by the FDA. Prominent among these were the Association of Black Cardiologists (ABC), the NAACP, and the National Minority Health Month Foundation (NMHMF). Each of these groups doubtless saw BiDil as an efficacious drug that promised to help their constituents. But it should also be noted that each also received substantial funding from NitroMed. The ABC received $200,000, the NMHF received an undisclosed amount of an "unrestricted educational grant to undertake epidemiological research on chronic heart failure patients," and NitroMed also provided $1.5 million to fund an NAACP "health justice" campaign.[11]

Significantly, however, many of the groups also argued that while BiDil should be approved, it should be approved for use *regardless* of race. Notably, the day before the FDA Advisory Committee meeting, the National Minority Health Month Foundation staged a press conference with an array of interest groups identified as being for African Americans including representatives from the Alliance of Minority Medical Associations; Association of Black Cardiologists; Genetic Alliance, Inc.; International Society on Hypertension in Blacks; Joint Center for Political and Economic Studies; Health Policy Institute; National Association for the Advancement of Colored People; and the National Medical Association. These groups issued a joint press statement calling for FDA approval of BiDil. The announcement garnered much media attention. Less noticed, however, was the fact that the press release contained such statements as "[t]he assertion that this is a race drug is misguided" and "[i]t would be 'bad science' to label or market this drug as a 'Black' drug. More importantly, race-based claims are not credible in the face of modern genetic science."[12] The press release itself was titled "Organizations Unite to Support BiDil's Approval for Heart Failure. Rebuff Designation as "Race-Only Drug." In the end, though, the FDA heard only NitroMed's voice and acceded to its request for a race-specific approval of BiDil.

CONCLUSION

The story of BiDil clearly raises concerns over the dangers of reifying race in a manner that could lead to new forms of discrimination. BiDil, however, is part of a much larger dynamic of reification in which the purported "reality of race" as genetic may be used to obscure the social reality of "racism." To the extent that this dynamic succeeds in reductively reconfiguring health and other types of disparity in terms of genetic difference, it casts personal responsibility and the market as the appropriate arenas for addressing differential outcomes. It also undermines the rationale for deliberate state or institutional interventions to address discrimination.

This is not to advocate "color blind" medicine. To the contrary, there are very real health disparities in the country that correlate with race. African Americans suffer a disproportionate burden of a number of diseases, including hypertension and diabetes. Like heart failure, these are complex conditions caused by an array of environmental, social, and economic as well as genetic factors. Central among these is the fact that African Americans experience discrimination, both in society at large and in the health care system specifically. The question is, once society identifies these disparities in health outcomes, how does it address the underlying causes? Of course, outcomes can have multiple causes, both social and genetic. But health disparities are not caused by an absence of "black" drugs. As studies by the Institute of Medicine, among others, make clear, they are caused by social discrimination and economic inequality. The problem with marketing race-specific drugs is that it becomes easier to ignore the social realities and focus on the molecules. For all the legitimate concerns that the genomics revolution might lead to new forms of discrimination, we must also be alert to the potential appropriation of genetics to obscure or justify existing inequalities.

NOTES

This chapter draws heavily on the following publications including two authored by myself: B. D. Smedley, A. Y. Stith, A. R. Nelson, eds., *Unequal Treatment: Confronting Racial and Ethnic Disparities in Health Care* (Washington, DC: National Academies Press, 2002); J. Kahn, "How a Drug Becomes 'Ethnic': Law, Commerce and the Production of Racial Categories in Medicine," *Yale Journal of Health Policy, Law & Ethics* 4 (2004): 1–46; J. Kahn, "Race in a Bottle," *Scientific American*, August 2007, 40–45; Sandra S. Lee et al., "The Meanings of 'Race' in the New Genomics: Implications for Health Disparities Research," *Yale Journal Health Policy Law & Ethics* 1 (2001): 33–75; A. Taylor et al., "Combination of Isosorbide Dinitrate and Hydralazine in Blacks with Heart Failure," *New England Journal of Medicine* 351 (2004): 2049–57.

1. Jonathan Kahn, "From Disparity to Difference: How Race Specific Medicines May Undermine Policies to Address Inequalities in Health Care," *Southern California Interdisciplinary Law Journal* 15 (2005): 105–29.

2. Richard Cooper et al., An International Comparative Study of Blood Pressure in Populations of European vs. African Descent," *BMC Medicine* 3 (January 5, 2005): 2, www.biomedcentral.com/1741-7015/3/2 (accessed March 9, 2010).
3. Editorial "Genes, Drugs and Race," *Nature Genetic* 29 (2001): 239.
4. Editorial, "Illuminating BiDil," *Nature Biotechnology* 23 (2005): 903.
5. Jay N. Cohn et al., "Effect of Vasodilator Therapy on Mortality in Chronic Congestive Heart Failure: Results of a Veterans Administration Cooperative Study," *New England Journal of Medicine* 314 (1986): 1547–52.
6. Jay N. Cohn et al., "A Comparison of Enalapril with Hydralazine-Isosorbide Dinitrate in the Treatment of Chronic Congestive Heart Failure," *New England Journal of Medicine* 325 (1991): 303.
7. J. Kahn. "How a Drug Becomes 'Ethnic.'"
8. Ibid., 31.
9. Peter Carson et al., "Racial Differences in Response to Therapy for Heart Failure: Analysis of the Vasodilator-Heart Failure Trials," *Journal of Cardiac Failure* 5 (1999): 178.
10. U.S. Patent #6,465,463, emphasis added.
11. Food and Drug Administration (FDA). Center for Drug Evaluation and Research, Cardiovascular and Renal Drug Advisory Committee. "Transcript of Meeting, June 16, 2005," 210–11, www.fda.gov/ohrms/dockets/ac/05/transcripts/2005–4145T2.pdf (accessed March 9, 2010); National Minority Health Month Foundation, "Organizations Unite to Support BiDil's Approval for Heart Failure. Rebuff Designation as "Race-Only Drug." June 15, 2005, www.redorbit.com/news/health/156280/organizations_unite_to_support_bidils_approval_for_heart_failure_rebuff/index.html (accessed March 9, 2010); TargetMarketNews, "NitroMed, NAACP Partnership Will Help Introduction of BiDil." December 14, 2005, www.targetmarketnews.com/storyid12150501.htm (accessed March 9, 2010).
12. National Minority Health Month Foundation, "Organizations Unite." The first statement was made by Randall W. Maxey, MD, President, Alliance of Minority Medical Associations; the second was made by Dr. Gail Christopher, Vice President of the Office of Health, Women, and Family, Joint Center for Political and Economic Studies.

8

EVOLUTIONARY VERSUS RACIAL MEDICINE

WHY IT MATTERS

Joseph L. Graves, Jr.

MAKING SENSE OF BIOLOGY AND MEDICINE

In the classic paper "Nothing in Biology Makes Sense Except in the Light of Evolution," Theodosius Dobzhansky explained why and how evolutionary biology was the core unifying theme in biology.[1] While medical practitioners do not always recognize this, medicine is a subset of biology, or at least successful medical intervention is heavily dependent upon biological principles. This fact was recognized as early as Erasmus Darwin in *Zoonomia, or the Laws of Organic Life*.[2] Thus, even before evolutionary theory took root, some physicians had intuitions that eventually led to the fully developed theory of evolutionary medicine.[3] Over the last two decades this field has been increasing in popularity. The new field of "Darwinian" or "Evolutionary" medicine was heralded with the publication of Neese and Williams's "Why We Get Sick" in 1994.[4] In this work, the authors argued that disease resulted from one or a combination of five basic causes: (1) Infection, (2) Novel Environments, (3) Genes, (4) Design Compromises, and (5) Evolutionary Legacies. The notion that disease results from infection and genetic variation was not a challenge to traditional medical thinking. However, the way that novel environment, design compromise, and evolutionary legacies contribute to disease requires some illustration. For example, modern humans were originally hunter-gatherers who lived on a limited calorie diet. Even with the advent of agriculture, most humans did not experience diets with consistent caloric excesses until recently. Thus, the increase of obesity-related diseases in the industrialized world results from the novel environment of excess calories. Design compromises occur due to anatomical, life history, and physiological trade-offs. For example, much of early human history was dominated by high infant and maternal mortality during child birth. The design compromise here is the size and shape of the human female pelvis compared to the size of the human infant's head. The mosaic feature of past human evolution developed an increased head size, while not simultaneously increasing pelvic width at the same rate. Thus, the greater mortality that resulted was due to a design compromise. Finally, evolutionary legacies result from the fact that evolution does not completely redesign structures, it adds on and modifies existing structures. Humans are chordates and show all of the chordate features at some stage of their life cycle. One features of chordates that contributes to human mortality is the fact that the windpipe and digestive tract cross each other. The epiglottis is an evolved feature that prevents food from passing into the respiratory tract. However, it doesn't always work, so choking, especially among human infants, has always been a concern.

Thus far, health disparity research and literature has not incorporated the full evolutionary medical approach. Generally, when biology is addressed as a cause of disparity and the focus has been on genetic differences that exist between reputed racial/ethnic groups, the evidence supporting the connection has been tenuous.[5] The logical errors concerning genetic causality result from either ignoring or misunderstanding evolutionary genetics. The lack of training in evolutionary biology among medical researchers and practitioners accounts for this oversight.[6] In particular, biomedical scientists often confuse the existence of geographically based genetic variation as proof of the existence of biological races. They also incorrectly assume that genetic differences in loci associated with complex diseases between populations is one of the causes of health disparity.[7]

CORE CLAIMS OF RACIAL MEDICINE

Racial medicine begins with the notion that biological races exist within modern humans. It also follows that meaningful racialized differences exist in the prevalence of diseases between these groups. Furthermore, racial medicine asserts that the predominant cause of these prevalence disparities is genetic differences. Thus, racial medicine would assert that we can reliably predict an individual's disease predisposition by determining what biological race they belong to. If this syllogism were true, its utility would be obvious. For example, we should be able to develop a number of race-specific therapies. This would allow pharmaceutical companies to market specific drugs to target races. BiDil was just such a drug. Originally designed as a better heart therapeutic for *all* individuals, it became racialized when it was shown that while the drug didn't significantly improve heart disease outcomes for the overall population tested, it did for the African American subsample in the study. As a result the manufacturer, NitroMed, sought and received permission from the FDA to market BiDil as the first "race-specific" drug. This right was granted despite the fact that BiDil was never extensively tested with other racial/ethnic groups. Indeed, the clinical trial showing that BiDil is a race-specific drug has significant flaws. The A-HeFT trial that propelled BiDil's FDA approval did not clearly support the claims of race specificity made by the drug's proponents. While the 43 percent reduction in mortality was a stunning feat, it is not sufficient to support the claim to race-specificity since it only enrolled self-identified blacks. Even Dr. Jay Cohn, the person who developed BiDil, acknowledged that nonblacks can receive a substantial benefit from the medication. Ironically, BiDil failed because African American patients for the most part did not desire a "race-tailored" drug.[8]

Despite the fact that most physicians practice medicine as if biological races are clearly defined in modern humans, there are few scientists or physicians who can or are willing to construct an argument to support racial medicine. The modern medical literature still utilizes nineteenth-century anthropological categories in group studies. For example, a recent search of Entrez Pubmed (conducted on January 6, 2010) utilizing the term "race" returned 114,305 articles from the human biomedical literature. More specifically, searches on Caucasian, Mongoloid, and Negroid race returned 52,846, 22,667, and 38,792 citations, respectively. While one can still debate the utility of the term "race" in the human biomedical literature, almost no one defends the idea that nineteenth-century racial categories are legitimate or that these (Caucasian, Mongoloid, or Negroid) are of much use in twenty-first-century research.

Amongst those who defend the notion that nineteenth-century racial conceptions have some utility is Satel.[9] In 2002, Dr. Sally Satel wrote in the *New York Times Magazine* that she was a "racially profiling doctor." Satel is a psychiatrist based in Washington, DC, a lecturer for Yale University, and W. H. Brady Fellow at the American Enterprise Institute. Satel argued that the biological basis of race was a clinical reality and that denying racial differences was dangerous for treatment.[10] However, more important than Satel's assertions are the arguments of Neil Risch, a professor of genetics at University of California, San Francisco.[11] Risch, along with his coworkers, has consistently argued that self-identified race should be used in epidemiological research. His claim is that self-identified race corresponds well to the actual biological races within our species. The utility of this claim follows from accepting the remainder of the racial medicine syllogism.

BIOLOGICAL RACE DEFINITIONS AND THE CLUSTERING GENETIC VARIATION

An example of this racial reasoning is illustrated by "Genetic Structure, Self-identified Race/Ethnicity, and Confounding in Case-control Association Studies," a study coauthored by Risch.[12] The study utilized 326 microsatellites of individuals who self-identified into particular racial/ethnic groups and determined how their genetic variation clustered. They examined genetic information from people who self-identified as white, black, Mexican American, Chinese American, Chinese (Taiwan), Japanese, and other. Microsatellites are one to four base pair sequences of DNA that are widely dispersed across the human genome (both in coding and noncoding segments.) They are particularly useful in identifying ancestry. The microsatellite clustering was achieved via the algorithm *Structure*, which utilizes multilocus genetic data to infer population structure and to assign individuals into populations.[13] The process of identifying populations within species is useful for solving a number of evolutionary and population genetic questions, such as determining population origins, identifying admixture, and evaluating subdivision. *Structure* can and is utilized to study these problems in a wide variety of species, and produces the number of clusters (K = 1, 2, 3, etc.) that the user specifies. Its default assumptions are that all genetic markers are unlinked, at linkage equilibrium with one another within populations, and at Hardy-Weinberg equilibrium within populations.[14] It also assumes that the genetic information of individuals is derived from a single population (cluster). The model can handle admixture, but that must be specified in the run.

The results of the study's clustering of self-identified ethnicity and genetic variation are shown in table 8.1. At first glance, one might get the impression that these data support the notion that biological races exist within the human species. The correspondence between genetically defined clusters (A, B, C, D) and self-identified groups is very high. However there are apparent problems in the way this analysis was run. The authors admit that they did not engage the admixture option in *Structure*. Given that two of the populations in question, US blacks and Mexican Americans are known to be admixed, this is a major error. Furthermore, the authors could have easily run the data set again with the admixture option enabled. This would have allowed them to compare the two treatments to determine if this second analysis gave significantly different results from the first. In the case of African Americans there is excellent indication that an admixture analysis would have shown quite different results. Another study using a much larger number of genetic markers (a full genome scan of about 250,000 SNPs) indicated that African American admixture ranged from as high as 99 percent to as low as 1 percent with a median value of 18.5.[15] Despite this apparent fault in the study procedure the authors were still confident that their analysis supported the notion that self-identified race corresponded to an underlying biological race structure within the human species.[16]

The largest misconception of this approach is that it ignores the fact that isolation by distance explains the vast majority of variation in human allele frequencies.[17] Thus, 75 percent of human allele frequency variation is explained by geographic distance.[18] This means that it is possible to

TABLE 8.1 SELF-IDENTIFIED RACE/ETHNICITY AND GENETIC VARIATION

SIRE*	Cluster A	Cluster B	Cluster C	Cluster D
Eu. Am.	1,348	0	0	1
Af. Am.	3	1,305	0	
Mex. Am.	1	0	0	411
Chin. Am. & Taiwanese	0	407	0	0
Japanese	0	160	0	0
Other	1	2	0	9

*Self-identified race/ethnicity

Source: H. Tang, T. Queltermous, B. Rodriguez et al., "Genetic Structure, Self-identified Race/Ethnicity, and Confounding in Case-control Association Studies," *American Journal of Human Genetics* 76 (2005): 268–75.

produce the appearance of clustering simply by where one samples genetic variation. Serre and Paabo demonstrated that heterogeneous sampling gave rise to genetic clusters that were biologically meaningless.[19]

This is precisely what is occurring in biomedical research in the United States. The fact that American studies discontinuously sample human genetic variation produces the appearance of human racial clustering. This occurs because of the social processes by which the American population came into existence. The five population groups examined in this study were self-identified Caucasians, African Americans, Hispanics (mainly Mexican Americans), Chinese Americans (with four Chinese grandparents), and Japanese Americans (with four Japanese grandparents). These groups are representative of the major population groups that inhabit the modern United States, but are in no way a comprehensive sample of the world's genetic diversity. Again, we know that human genetic variation is associated with geographic distance.[20] For example, Tishkoff et al. showed that *Structure* could cluster five groups of African populations by region (Saharan, Western, Central, Eastern, and Southern).[21] This analysis simply states that these populations share more polymorphic genetic variants with each other, not that they are biological races. Thus, it was not surprising that the cluster analysis utilized in

"Genetic Structure" using far fewer genetic markers could separate these groups (whose ancestry is separated by much greater genetic distances than those which occur on the African continent).

BIOLOGICAL DEFINITIONS OF RACE

Some proponents of racial medicine have actually outlined what they mean by biological race within the human species and a line of reasoning that might support the notion that anatomically modern humans show such variation.[22] There have been four ways in which the race concept has been conceived: essentialist, taxonomic, population, and lineage. Essentialist notions of race are ancient, and are not necessarily related to any concept of common descent. Essentialism claims that there is an essence of traits that can be understood as characterizing a species (and thus any races that might exist within it). The taxonomic race concept suggests that there are aggregate populations of a species possessing phenotypic similarities and inhabiting geographic subdivisions of the range of the species.[23] Of these, evolutionary biologists long ago rejected the essentialist and taxonomic definitions of species (and thereby those same definitions applied to race as well). This means that only the population and lineage definitions of race retain any utility.

Population definitions of race revolve around how much genetic variation exists within and between supposed racial groups. If there is more genetic variation within a group than between them, we really cannot support the notion that the groups have diverged sufficiently to describe biological races. For evolutionary biology, biological races occur as part of the speciation process. Through the species' history, local adaptation and genetic drift may cause sufficient changes in some populations such that they eventually form new species from the original founders. This process is dynamic, and populations may diverge or converge genetically over the lifetime of the species without ever giving rise to new ones.

Obviously, given this process, the amount of variation cannot be evaluated at one or a small number of genetic loci. Or so it would seem if geographical races result from the gradual process of speciation. Ernst Mayr discussed this in his classic work, *Population Species, and Evolution.*[24]

> Biological races. A necessary corollary of any theory of gradual speciation is that there should exist in nature "forms" or "varieties" or "populations" that are incipient species. . . . Kinds of animals that show no (or only slight) structural differences, although clearly separable by biological characters, are called biological races.

Mayr did not explain what he meant by "clearly separable," but he did state that of all the phenomena listed as races in the biological literature, the best case for the existence of true biological races in nature are the host races formed by various insect species on different plant species. It is notable that he did not think that humans fit this definition well. Yet it does not take much genetic variation at all to account for host-race formation in insects. In the most well-studied case of this, *Rhagoletis pomonella* (the apple maggot fly) has formed two host races throughout its geographic range. The fly ancestral host is the hawthorn tree (*Crataegus* spp.) About 150 years ago, this fly species was recorded as being a pest of cultivated apple trees (*Malus*). It is now known that allele frequencies at about four loci are differentiated between the two host races.[25] In this case, the loci deal with simple traits that impact fitness on the host plants, feeding and development time, and mate choice. Thus, in the case of insect host races (which are biological races), one does not need many loci to create races. No one, however, would claim that racial formation in organisms with more complex behavior is this simple.

POPULATION SUBDIVISION

Another study examined 4,833 SNPs in 538 clusters across the human genome in European (N = 30 individuals), African Americans (N = 30 individuals), and Asians (N = 40 individuals).[26] This study evaluated the population subdivision statistic (FST), which compares the allelic diversity of each of the subpopulations against a pooled total population. Since Sewall Wright's invention of F coefficients, which examine the proportioning of genetic variation between different levels within a species, population geneticists have utilized a minimum value of differentiation between subpopulations and the total species as the threshold for identifying the existence of biological races (FST > 0.250).[27] In this study, the mean frequency for FST at each locus was 0.083, with only 10 percent of the loci exceeding FST of 0.18 and about 6.5 percent exceeding FST of 0.250. This is consistent with the general finding that, averaged across the genome, FST in humans does not approach Wright's threshold (and is generally FST = 0.11).

Woodley disputes the notion that population subdivision is a sufficient reason to invalidate the existence of race in anatomically modern humans. This is first attempted by reference to "Lewontin's fallacy." Lewontin's fallacy was first coined by the statistician A. W. F. Edwards in an essay that appeared in the journal *Bioessays* in 2003.[28] This piece claimed that population geneticist Richard Lewontin was in error when he argued that the amount of genetic variation within populations was far greater than that between populations, which therefore invalidated the ability to assign individuals to racial groups. Edwards argued that if enough genetic loci are used, individuals can be distinguished into clusters, and that such clusters could represent biological races. Graves points out the many problems with this analysis.[29] The largest problem is how population genetic data have been collected thus far. Generally, samples are taken from groups who are descended from populations whose geographic range is very far apart. Thus, we expect clustering of allele frequencies if one compares sub-Saharan Africans to Northern Europeans to East Asians. However, if the full continuum of human populations were examined, when a specific cluster began and when another ended would not be so clear. For example, in 2009 54,794 SNPs were examined in 1,928 individuals from 73 Asian populations. These data were compared with data from sub-Saharan Africans (Yorubans) and European Americans. Running *Structure* with K = 14 showed that linguistic groups tended to cluster together; however, there were populations that fell in clusters that did not belong to their linguistic or geographic affinity.[30]

CLADISTIC RACES

The other criterion by which biological races might be identified is their unique genetic lineages.[31] Andreasen argues that is precisely how human races can and should be identified (by cladistic races). Thus, if the human species really has unique genetic lineages, then it should be possible to represent various human races on an evolutionary tree. Templeton pointed out that if this were so, then the genetic distances between all non-African and African populations should be the same.[32] If on the other hand, genetic distances between populations reflect the amount of gene flow, then there should be a strong relationship between geographic and genetic distance, which is referred to as "isolation by distance" (IBD).[33] But we do not find that the genetic distances between all non-African and African populations are the same, and thus the cladistic race argument fails immediately. On the other hand, we do find that genetic distance (measured by population subdivision, FST) and pair-wise geographic distance between populations shows a highly statistically significant correlation ($R2 = 0.7679$).[34] Despite the overwhelming use of cluster algorithms (*Structure*) to examine human diversity (which is consistent with the notion of independent evolutionary lineages), the data strongly supports the IBD explanation

RACIALIST MISCONCEPTIONS

There are severe consequences to getting the apportioning of human genetic variation incorrect in medicine. Incorrect thinking about race has important implications for the way biomedical research studies are structured and their data collected and evaluated, and on the public policy proscriptions that result from this process. For example, discordance is a major problem for the notion of racial medicine.[35] Discordance results from the fact that NS has acted on different portions of the human genome that are unlinked. Thus, selection for particular skin color variants need not have any relationship to selection for resistance to a particular pathogen. For example, genetic variation at the vitamin D binding locus is clinal following solar intensity from the equator to northern and southern latitudes.[36] The vitamin D binding locus is found on chromosome 12. The major action of vitamin D is to stimulate absorption of calcium in the intestine. Because calcium takes part in a wide variety of cellular processes, genetic variation at this locus has pleiotropic impact on the phenotype. Pleiotropy refers to a genetic locus that has multiple impacts on many physiological systems. Another pleitotropic locus is SM1, which is found at chromosome

5q31-q33. This locus regulates the immune response to Schistosome parasites. Approximately three hundred million people are currently exposed to schistosome parasites, with infections concentrated in Africa, Asia, and South America. SM1 is a quantitative trait locus that includes genes involved in the regulation of immune responses to pathogens, such as the interleukins (IL-4, IL-5, IL-9, and IL13).[37] This region of the chromosome also includes loci that impact the proinflammatory cytokines (IL12 and IL23), the interferon regulatory factor (IRF1), colony stimulating factor (CSF1R), regulation of IgE levels, familial eosinophilia, asthma, *Plasmodium falciparum* (malaria) infection levels, and inflammatory bowel disease. Variation at SM1 is not clinal, it is related to the presence of and exposure to *Schistosoma* spp. parasites. For example, the distribution of *S. haematobium* in Africa shows populations within the same latitude that are heavily exposed, while others are mildly or not exposed.[38] Thus, the relationship between aspects of the phenotype that result from any genetic variation at vitamin D binding and SM1 loci will not be consistent within or between populations. Given the complexity that results from just these two loci that impact so many physiological systems, how could we ever expect to develop a "racial" medicine?

Furthermore, for racial medicine to have utility, one would have to first be able to identify human races and, second, there would have to be a high certainty that individuals within these races would share specific disease-related genes in common that are different from those of other races. This scenario fails at even the simplest levels, especially within the United States. One study found that the genetic ancestry of African Americans varied widely (averaging 0.69–0.74 alleles most common in Western Africa and 0.11–0.15 alleles most common in Europe and the Middle East).[39] A later study found that the median reputed European ancestry of self-reported African Americans was 0.185 but that the subjects varied in frequency from 0.99 to 0.01 European American.[40] Thus, for this population the notion of racial medicine is absurd from the outset. This of course begs the question of how BiDil could have ever been approved for African Americans as the first "race-specific" drug.

THE GENETICALLY SICK AFRICAN AMERICAN

Ironically, so much of the justification for pursuing "racial" medicine resulted from the ongoing and persistent health disparities observed in African Americans. Throughout the twentieth century, age-specific mortality resulting from a variety of biologically based diseases for African Americans was double that of European Americans.[41] In the context of what we know about human genetic variation, this result really cannot be explained genetically. Yet, this is precisely the tenor of the genetic health disparity research program. A case in point: a study argued that a more active version of a gene leukotriene A4 hydrolase was responsible for contributing to greater heart-attack risk in African Americans compared to European Americans.[42] The gene is involved in the synthesis of leukotrienes, agents that maintain a state of inflammation, which helps regulate the body's inflammatory response to infection. The authors of the study argued that the more active version of this gene might have risen to prominence in Europeans and Asians because it conferred extra protection against infectious disease (via positive natural selection [NS]). Thus individuals with this genetic variant would acquire protection from disease early in life but later in life would have a higher risk of heart attack (negative selection or NS–). Without realizing this, the authors were arguing that the frequency of this leukotriene A4 hydrolase allele was governed by antagonistic pleiotropy.[43] This would occur because plaques that build up in the walls of the arteries could become inflamed and rupture. The increased risk of heart attack in Europeans and Asians had diminished through time because the active version of the gene was favored in them long ago. These populations therefore had time to evolve genetic changes that offset the extra risk of heart attack. Conversely, since the active version of the inflammatory gene passed from Europeans into African Americans only a few generations ago (about ten generations), this would have been too short a time to develop genes in the AAs that protected against heart attack. Graves explained why this scenario could not possibly be true.[44] It fails because it ignored the actual distribution of leukotriene A4 hydrolase allele (cosmopolitan), ignored the potential of linkage between the locus in question and the modifier, ignored the fact that heart attack was not a significant source of death until later in the twentieth century, and did not recognize that NS is generally unconcerned with positive fitness impacts late in life. This example is important because it illustrated the general ideological bent of modern biomedical research. That is, if a health disparity exists and a genetic difference can be found, the genetic difference must be responsible for the health disparity.

A 2009 study presented another example of this general fallacy (namely, false causation).[45] The authors of this study concluded, like so many others, that because specific genetic variants are found at highest frequency in persons of non-European descent and because these people have the highest frequencies of late-age complex disease, it must be the genes that are playing the most significant role. This argument is made even more absurd by recent genome-wide studies which show that Europeans actually have a greater load of deleterious mutations than Africans do.[46] Therefore, the real question is, if more of the European alleles are deleterious, why are European Americans living so well and African Americans living with so much disease?

EVOLUTIONARY MECHANISMS OF DISEASE, RACE, AND HEALTH DISPARITY

The main premise of evolutionary medicine is that disease results from combinations of infection, novel environments, genes, design compromises, and evolutionary legacies. If we are concerned with health disparity between socially defined racial groups, design compromises and evolutionary legacies probably are not major players. These are not candidates for creating health disparities because these should be shared among all humans. Examples of a design compromise include standing upright or neonatal heads passing through the mother's pelvic girdle. An evolutionary legacy is some aspect of our anatomy or physiology which results from our past species ancestry. Thus, the fact that our trachea and esophagus cross results from our ancestry as chordates.

Design compromise or evolutionary legacy could contribute to health disparity in this context if combined with one of the other three sources, infection, novel environment, or genetic differences. However, genes that have large impacts on health and reproduction will necessarily be rare via mutation/selection balance. Such genes will equilibrate around their mutation rate and in geographically separated populations their frequencies will be determined by genetic drift. Thus, when we think of racialized health disparity, we are not thinking about syndromes like progeria, fragile X syndrome, phenylketonuria, Crohn's, or even Tay-Sachs disease. The diseases that have played into the discussion of health disparity in the United States include cancer, stroke, heart disease, diabetes, and even aging itself. These diseases are influenced by a large number of genes, but they all have significant environmental inputs as well. Thus, to understand how to approach eliminating racialized health disparity, we really must utilize the full intellectual arsenal of evolutionary genetics. Thus far, this simply has not happened.[47]

Because mutations are rare, $\sim 1 \times 10^{-4}$ to 1×10^{-6} per replication of DNA, mutation alone could not possibly significantly change the genetic composition of populations. Thus, changes in gene frequency occur via the mechanisms of natural selection (differential reproduction of favored genotypes) and genetic drift (random changes that result from fluctuation in population size through time). With regard to complex human diseases, it is important to understand that the action of natural selection is age-specific.[48] Mutations that have deleterious impacts on early reproduction face strong negative natural selection (NS–) and will become very rare. The expected frequency of these traits over time is equal to the rate at which they appear by mutation (μ, this is called mutation/selection balance. For example, progeria is a genetic disease that gives the appearance of premature aging. Progeriacs rarely live beyond thirteen and almost never reproduce. Thus, by mutation/selection balance, the frequency of this trait is between 1 in 4 to 8 million worldwide or between 2.5×10^{-7} to 1.25×10^{-7}. Conversely, traits whose impact occurs after an individual's net reproductive value is zero (at about age 40) face little NS (–). These traits will accumulate in populations (mutation accumulation) and their frequency will be determined by chance (genetic drift).

Disease can result from acute or stealth infection (which is common), inborn errors of metabolism (rare according to the argument above), and the unresolved contingencies of our evolutionary heritage (again, common). For example, most of the diseases associated with health disparity (cancer, heart disease, stroke, diabetes) all show strong age-specific profiles. As such they can be seen as diseases of aging. However, aging itself is an evolutionary contingency.[49] The evolutionary theory of aging tells us that populations accumulate alleles that allow aging via two mechanisms, mutation accumulation (genetic drift on genes that are neutral in early life, but cause disease in late life) and antagonistic pleiotropy (genes that are beneficial in early life, but cause disease in late life).[50] An example of a disease whose frequency is determined by mutation accumulation is Alzheimer's disease. At present we know of no negative fitness that results from the alleles associated with Alzheimer's at early age. Thus, its prevalence in relatively large populations worldwide will be governed by past population history, and in smaller populations the frequencies will be somewhat random. Studies of the frequency of the epsilon 4 allele at the apolipoprotein locus (<ΣYB>ε</ΣYB>4 is associated with greater Alzheimer's risk) validate this claim. In Europe, the frequencies of <ΣYB>ε</ΣYB>4 in large samples are close to 0.110–0.150, in Chinese samples 0.049–0.081, Amerindian samples 0.112–0.193, and sub-Saharan African and African American samples about 0.250. Consistent with this claim is the variation that occurs in <ΣYB>ε</ΣYB>4 frequency when groups that live in small populations are sampled. The frequency for the Khoi-San of Southern Africa was 0.370 and five Brazilian Amerindian groups ranged from 0.043 to 0.423.[51] It is notable that the Marin study could not detect the presence of the ε2 allele in the Brazilian Amerindian groups they examined. The loss of alleles routinely occurs due to genetic drift in small populations.[52]

Few researchers understand the implications of the evolutionary theory of aging and its now well-established experimental verification for understanding human disease and health disparity. First, antagonistic pleiotropy tells us that there are genes that contribute to disease at later age that are present in all populations. The ubiquity of cancer genes in the mammalian genome is an example of selection in deep evolutionary history for genes essential to mammalian development. However, in modern humans, these genes contribute to late-life disease and are therefore an evolutionary legacy.[53] Therefore, all human populations have genes at high frequency that contribute to cancer. The logical implication of this is that if we see disparities in cancer prevalence and mortality, we need to examine other mechanisms. For example, in the United States stomach cancer is found at higher incidences amongst East Asians (Japanese, Chinese, and Koreans) compared to African and

European Americans.[54] However, human genetic factors are not primary factors with regard to this disparity. The risk factors identified include *Heliobacter pylori* infection, tobacco smoking, high salt intake, low intake of fruits and vegetables, high intake of pickled foods, high intake of nitrates and nitrate-related compounds, and family history (genetic). Indeed, a 2009 study indicates that genetic variation in the *Heliobacter pylori*, specifically the CagA-positive strains, are highly associated with noncardiac gastric carcinoma in patients from Taiwan. In Taiwan, 95 percent of *H. pylori* strains are CagA-positive.[55] This example demonstrates that you cannot assume that any cancer disparity results from a genetic difference between the human populations you observe, since so many other factors are at play.

Mutation accumulation can also contribute to the genetics late-life disease disparities, since these allele frequency differences result from past population history and random fluctuations resulting from small population size. Again, the significance of this fact has not been fully comprehended by biomedical researchers. The lower allelic diversity of Europeans and Asians compared to sub-Saharan Africans resulted from genetic drift due to a past population bottleneck that resulted from the migration of the former into their present geographic ranges out of Africa.[56] The notion that this process led to the greater genetic hardiness of these groups compared to sub-Saharan Africans is never stated. Yet, this conclusion is inescapable when one examines the plethora of research studies that claim that the disease disparity of African Americans in North America results in the main from their African genetic heritage.[57] This thinking is well illustrated in a study by Chang et al.[58] This study examined selected candidate gene variants in the US populations reputedly associated with complex disease. The variants were selected from genes that are known to play a role in six major cellular and physiological pathways: nutrient metabolism, immune and inflammatory responses, xenobiotic mechanisms, DNA repair, hemostasis and the renin-angiotensin-aldosterone system, and oxidative stress. They found that significant differences occurred at 87/90 genetic loci examined in the study by racial/ethnic group (non-Hispanic white, non-Hispanic black, and Hispanics). They concluded, not surprisingly, that differences in gene frequencies may be contributing to health disparities in the disease that result from problems in these pathways.

While the evolutionary mechanisms that account for genetic diversity are well understood, the process by which genes produce phenotypes is not. For example, Giot et al. have examined the protein map in *Drosophila melanogaster*.[59] This study found 3522 proteins with 3,000 interactions that were orthologous to human disease. Orthologous genes are those that share sequence homology due to a speciation event. Thus, these genes would have been present in the last common ancestor of the Arthropod and Chordate line. The complexity of this is revealed when we consider that the average size of a human protein is about 476 amino acids. Each amino acid is coded by 3 nucleotides in sequence in the DNA code; thus, for the average protein there will be 1,328 nucleotides. To understand disease in *Drosophila,* we have to deal with potential variation occurring in at least 4.67 million nucleotides. This number is undoubtedly higher in humans.

Further complicating this picture is the most recent notion about the general weakness of genome-wide association studies to localize genetic variants that account for complex disease.[60] For example, for a well-known complex trait of high heritability (height) 20 SNPs explain about 3 in the variation and to account for 80 percent of the variation about 93,000 SNP's would be required.[61] Imagine how difficult the analysis becomes for even more complicated traits such as longevity. Lifespan differentials between African Americans and European Americans have been observed since the onset of chattel slavery in the Americas.[62] This begs the question of whether this differential resides within genetic variants associated with complex disease, which ultimately cause death. This question can be addressed by examining the heritability of life span, or the degree to which offspring resemble their parents in a complex trait due to genetics (h^2, which can reside between 0 and 1.00). Finch and Tanzi suggest that the heritability of human life span is about 0.23–0.35.[63] This would mean that 23 to 35 percent of the variation in human life span is accounted for by genetic effects, while the remainder (75 to 65 percent) would be accounted for by environment, gene x environment interaction, covariance of genes and environment, and any error that occurred in the measurement of the trait. Studies of the heritability of human life span by racial/ethnic group are sparse.[64] The best studies come from the Scandinavian countries (Denmark, Finland, and Sweden) where excellent records have been kept on monozygotic and dizygotic twins. These studies have generally concluded that h^2 for longevity is about 0.20–0.30.[65] However, one study found that the h^2 for human life span in the last century in New York City was 0.30, 0.15, and 0.10 for European Americans, Caribbean Hispanics, and African Americans, respectively.[66] All of these studies taken together indicate that at least 70 to 90 percent of the variation in human life span is nongenetic. They also reveal a perplexing disparity: the h2 of life span in African Americans in New York City was less than that of European Americans, indicating that more nongenetic effects were at work. In other words, during the twentieth century in New York City, African Americans experienced environments that further reduced their hypothetical life spans more so than European Americans. Another way to understand this is to consider that in 1950, African American life spans were about 87 percent of European American life spans.[67] If we apply the heritability estimates to the entire nation, then African American environmental effects on life span were three times higher than those of European Americans. Given the history of the United States, there would have been no reason to believe that in this time period the environmental impacts on African Americans were mainly positive. This comparison suggests that one can easily explain the racial/ethnic health disparity in the United States without any reliance on genetic differences.

CONCLUSION: RACIAL MEDICINE AS A RED HERRING

The first truly scientific analysis of human biological diversity began with Charles Darwin about 150 years ago. Darwin's greatest fear was his society's reaction to the notion that all human beings shared common ancestry with one another as well as with the rest of organic life.[68] It would take more than a hundred years before the scientific community began to accept the notion of human relatedness and to begin amassing genetic data quantifying the degree of human relatedness. What we have learned since then is that anatomically modern humans are a young species. The first agreed-upon modern human fossils are dated to about 120,000 years before present (ybp) and appear in Tanzania. The first modern human fossils outside of Africa are dated to about 100,000 ybp and appear in Israel. Today, the genetic evidence also suggests that all modern humans are descended from ancestors that once lived in sub-Saharan Africa. This has been shown by data derived from nuclear, mitochondrial, and Y chromosome DNA. It also follows that because humans are a young species, we don't have a great deal of genetic variation compared to other large-bodied mammals. For example, chimpanzees from three different regions of Africa (Eastern, Western, and Southern) have 43.33 times more genetic variability between them than the most genetically different human populations do.[69]

The small amount of genetic variation within our species explains why we don't show biological races under Ernst Mayr's notion of incipient species. Mayr thought that in some cases natural selection and genetic drift produce incipient species that are akin to subspecies or geographical races. However, the small amount of genetic variation that exists in humans is inconsistent with this stage of speciation. On the other hand, it is clear that human populations do have geographically based genetic variation. What is not clear is why any professional scientist should believe that these genetic variants explain the differentials for complex disease we see in American society. Graves and Rose argue that while human populations do differ in the frequency of genetic variants, the evidence associating this variation with the differentials we observe in the prevalence of complex disease is extremely weak.[70] How can it be that, while the body of evolutionary theory rails against the notion of genetically based differentials in complex disease among American racial/ethnic groups, this proposition is still so popular? To explain this contradiction, we must look to ideology and the social-political practices of the United States, not biology.

Few would argue against that African Americans have higher frequencies of the sickle-cell anemia variant than Europeans of Western European ancestry. It has been shown that persons who are heterozygous at this locus have elevated protection against the malaria parasite. Thus, if a group's health status was always about having the correct genes, we should never have expected African Americans to have higher mortality rates from malaria than European Americans. Yet, between the years 1921 and 1923, ten Southern states reported that the malaria death rate for African Americans was three times greater than for European Americans (25 per 100,000 versus only 7 per 100,000).[71] Indeed, in the same time period, African Americans suffered disproportionate death rates due to pellagra (a vitamin-deficiency disease), while Charles Davenport and the

Eugenics Record office railed against the victims of this disease and their many genetic deficiencies.[72]

These sad episodes illustrate a simple fact of public health. Mortality and morbidity have always been strongly influenced by social conditions. For example, in the 1930s, African Americans suffered elevated rates of homicide by European American lynch mob violence, while today the African American homicide rate is still elevated but now due to "black on black" crime.[73] In the former case, a genetic research program to evaluate the causes of the homicide disparity would have focused on genes controlling violent behavior in Europeans, while the latter would have focused on such genes in Africans. Both research programs would have been a fool's errands.

Modern researchers generally understand that all complex diseases have both genetic and nongenetic risk factors. Case in point, the h^2 of diabetes is between 0.87 to 0.28.[74] The higher estimate suggests that there is little that can be done by modifying environment (diet and exercise) while the latter estimate shows that such modifications would be quite powerful. The differences in h^2 estimates may differ due to variation in the gene loci responsible for causing the disease in different families. This is a question for ongoing genetic research. However, in the last two decades, the number of people in the United States with physician-diagnosed diabetes doubled.[75] This clearly cannot result from massive changes in the US gene pool in that same time period. During this time period, it has been shown that race, sex, obesity, and age are consistent predictors of the prevalence of diabetes.[76] With regard to race, the question must be asked, is it the genetic or the social aspect (or their interaction) that account for the consistent link between the race variable and complex diseases? Thus far, health disparity research has been driven by the assumption that the genetic aspect of race is important if not the most important, yet there is absolutely no reason to believe this. For example, Kouznetsova et al. found that residential proximity to hazardous waste sites causes a statistically significantly higher rate of hospitalization for diabetes (1.25 to 1.36 times).[77] The obvious connection is that persons living close to hazardous waste sites are exposed to greater levels of persistent organic pollutants (dioxins, furans, PCBs, and chlorinated pesticides). We know that there is an association between socially constructed race and the likelihood that one lives near a toxic waste site.[78] Thus, it is true that in the United States, a non–European American is highly likely to be exposed to environmental toxins that contribute to their increased likelihood to develop a complex disease in later life.

At present, the National Institutes of Health is devoting $2.7 billion per year to health disparity research.[79] There is no simple way to determine how much of these funds are devoted to projects that are organized in ways that will not produce results that are useful to eliminating or reducing health disparities. However, the research literature addressing this subject is not encouraging. This is because the operational assumption behind much of health disparities research is focused on how the genetic predisposition of ethnic minority populations contributes to their acquiring complex disease, as opposed to how the social dominance of ethnic majority populations is the causative agent of the diseases of ethnic minorities. Until this false paradigm is toppled, I submit that much of this research is following a fool's errand.

NOTES

1. Theodosius Dobzhansky, "Nothing in Biology Makes Sense Except in the Light of Evolution," *The Ameircan Biology Teacher* 35 (1973): 125–29.
2. Erasmus Darwin, *Zoonomia: or the Laws of Organic Life In Three Parts* (1796; Cambridge: Cambridge University Press, 2009).
3. W. R. Trevathan, E. C. Smith, and J. J. McKenna, *Evolutionary Medicine and Health*, (New York: Oxford University Press, 2008).
4. R. Neese, and G. C Williams, *Why We Get Sick: The New Science of Darwinian Medicine* (New York: Vintage, 1996).
5. T. LaVeist, *Minority Populations and Health: An Introduction to Health Disparities in the United States* (San Francisco, CA: Josey Bass, 2005), 24; J. H. Fujimura, T. Duster, and R. Rajagopalan, "Race, Genetics, and Disease: Questions of Evidence, Matters of Consequence," *Social Studies of Science* 38, no. 5 (2008): 643–56; J. L. Graves, *The Race Myth: Why We Pretend Race Exists in America* (New York: Dutton, 2005); J. L. Graves, "Biological V. Social Definitions of Race: Implications for Modern Biomedical Research," *Review of Black Political Economy* 37, no. 1 (2009): 43–60; DOI: 10.1007/s12114-009-9053-3.
6. Trevathan et al., *Evolutionary Medicine and Health*.
7. Graves, *The Race Myth*; Graves, "Biological V. Social Definitions of Race."
8. P. Sankar and J. Kahn, "BiDil: Race Medicine or Race Marketing?" *Health Affairs*, October 11, 2005, DOI: 10.1377/hlthaff.w5.455, http://content.healthaffairs.org/cgi/content/long/hlthaff.w5.455/DC1; J. Kahn, "Race-ing Patents/Patenting Race: An

Emerging Political Geography of Intellectual Property in Biotechnology," *Iowa Law Review* 92 (2007): 355–415.

9. Sally Satel, "I Am a Racially Profiling Doctor," *New York Times*, May 5, 2002.
10. Ibid.
11. Neil Risch, Esteban Burchard, Elad Ziv, and Hua Tang, "Categorization of Humans in Biomedical Research: Genes, Race, and Disease, *Genome Biology* 3, no. 7 (2002): 1–12, http://genomebiology.com/2002/3/7/comment/2007; Neil Risch, "Dissecting Racial and Ethnic Differences," *New England Journal of Medicine* 354, no. 4: 408–11.
12. H. Tang, T. Queltermous, B. Rodriguez et al., "Genetic Structure, Self-identified Race/Ethnicity, and Confounding in Case-control Association Studies," *American Journal of Human Genetics* 76 (2005): 268–75.
13. J. K. Pritchard, M. Stephens, and P. Donnelly, "Inference of Population Structure Using Multilocus Genotype Data," *Genetics* 155 (2000): 945–59.
14. Genes are unlinked if they do not occur on the same chromosome, or if they occur on the same chromosome they are far enough apart that they will assort independently during gamete formation (meiosis). Linkage equilibrium addresses whether two genetic markers have no impact on each other's frequency (via natural selection). If different allele combinations at the two genetic loci do impact their frequency, we call this linkage disequilibrium. Finally, the Hardy-Weinberg equilibrium refers to the expected genotype frequencies any locus is expected to show if populations mate at random, are large in size, and have no natural selection impacting the frequency of alleles at that locus. The expected genotypes are simply the binomial expectation of all combinations of the alleles.
15. K. Bryc, A. Auton, M. R. Nelson et al., "Genome-wide Patterns of Population Structure and Admixture in West Africans and African Americans," *Proceedings of the National Academy of Sciences* 107, no. 2: 786–91, www.pnas.org/cgi/doi/10.1073/pnas.0909559107.
16. Ibid.
17. A. R. Templeton, "The Genetic and Evolutionary Significance of Human Races," in *Race and Intelligence: Separating Science From Myth*, ed. J. Fish (Mahwah, NJ: Laurence Earlbaum, 2002), 31–56; L. J. Handley, A. Manica, J. Goudet, and F. Balloux, "Going the Distance: Human Genetics in a Clinal World," *Trends in Genetics* 23, no. 9: 432–39.
18. S. Ramachandran, O. Deshpande, C. Roseman et al., "Support from the Relationship of Genetic and Geographic Distance in Human Populations for a Serial Founder Effect Originating in Africa," *Proceedings of the National Academy of Sciences* 102 (2005): 15942–47; F. Prugnolle, A. Manica, and F. Balloux, "Geography Predicts Neutral Genetic Diversity of Human Populations," *Current Biology* 15 (2005): R159–R160; B. Linz, F. Balloux, Y. Moodley et al., "An African Origin for the Intimate Association Between Humans and *Heliobacter pylori*," *Nature* 445 (2007): 915–18; Handley, "Going the Distance."
19. D. Serre and S. P. Paabo, "Evidence for Gradients of Human Genetic Diversity Within and Among Continents," *Genome Research* 14 (2004): 1679–85.

20. Ramachandran et al., "Support from the Relationship of Genetic and Geographic Distance"; Prugnolle et al., "Geography Predicts Neutral Genetic Diversity"; Linz et al., "An African Origin"; Handley, "Going the Distance."
21. S. A. Tishkoff, F. A. Reed, F. R. Friedlaender et al., "The Genetic Structure and History of Africans and African Americans," *Science* 324 (May 22, 2009): 1035–44.
22. R. O. Andreasen, "A New Perspective on the Race Debate," *British Journal for the Philosophy of Science* 49 (1998): 199–225; R. O. Andreasen, "The Cladistic Race Concept: A Defense," *Biology and Philosophy* 19 (2004): 425–42; M. A. Woodley, "Is *Homo sapiens* Polytypic? Human Taxonomic Diversity and Its Implications," *Medical Hypotheses* 74 (2010): 195–201.
23. E. Mayr, *Principles of Systematic Zoology* (New York: McGraw Hill, 1969); Woodley, "Is *Homo sapiens* Polytypic?"
24. E. Mayr, *Population, Species, and Evolution* (Cambridge, MA: Harvard University Press, 1974), 258–59.
25. J. L. Feder and A. A. Forbes, "Host Fruit-odor Discrimination and Sympatric Host-race Formation," in *Specialization, Speciation, and Radiation: The Evolutionary Biology of Herbivorous Insects*, ed. K. J. Tilmon (Berkeley: University of California Press, 2008); A. P. Michel, J. Rull, M. Aluja, et al., "The Genetic Structure of Hawthorn-infesting *Rhagoletis pomonella* Populations in Mexico: Implications for Sympatric Speciation," *Molecular Ecology* 16 (2007): 2867–78.
26. A. G. Clark, R. Nielsen, J. Signorovitch et al., "Linkage Disequilibrium and Inference of Ancestral Recombination in 538 Single-nucleotide Polymorphism Clusters Across the Human Genome," *American Journal of Human Genetics* 73 (2003): 285–300.
27. See S. Wright, *Evolution and the Genetics of Populations,* vol. 4: *Variability Within and Among Natural Populations* (Chicago, IL: University of Chicago Press, 1978), 449–50; Templeton, "The Genetic and Evolutionary Significance of Human Races"; Graves, *The Myth Race*; Graves, "Biological V. Social Definitions of Race."
28. A. W. F. Edwards, "Human Genetic Diversity: Lewontin's Fallacy," *Bioessays* 25 (2003): 798–801.
29. Graves, *The Myth Race.*
30. HUGO Pan-Asian SNP Consortium 2009, "Mapping Human Genetic Diversity in Asia, *Science* 326 (December 11, 2009): 1541–45. Woodley ("Is *Homo sapiens* Polytypic?") also makes the error of equating a species's overall heterozygosity with the number of subspecies (races) that have been traditionally assigned to the species. Thus, he finds it contradictory that chimpanzees have an average heterozygosity of 0.63–0.73 and bonobos of 0.48 and each has four subspecies named, while anatomically modern humans have greater heterozygosity at 0.776 but only one subspecies named. The problem with this analysis is that the process of speciation (by which geographic races would develop) does not proceed by overall heterozygosity, but rather by the differences between the heterozygosity of subpopulations from the total heterozygosity of the species: FST = (HT–HS)/HT, where this FST is the average for multiple loci; HT is the average of the expected heterozygosity in the total population over loci, and HS is the average expected heterozygosity over

subpopulations (P. W. Hedrick, *Genetics of Populations*, 3rd ed. [Sudbury, MA: Jones and Bartlett, 2005]). Thus, geographical races form as subpopulations and begin to diverge from each other in allele frequencies. A species might have very high heterozygosity, but if it is evenly distributed across its populations, then it isn't undergoing genetic change that will form geographical races that might eventually form new species.

Woodley partially addresses his misconception concerning average heterozygosity when he cites FST between sub-Saharan African and Australian Aborigine SNPs at FST = 0.33, based on data from Salter (F. K. Salter, *On Genetic Interests: Family, Ethnicity, and Humanity in the Age of Mass Migration* [Frankfurt: Peter Lang, 2003]). This value exceeds Wright's threshold compared to other species-level differences such as modern human versus Neanderthal mtDNA (FST = 0.08); Western Gorilla versus Eastern Gorilla mtDNA (FST = 0.02–0.09); Western Gorilla versus Eastern Gorilla nuclear loci (FST = 0.38); Common Chimpanzee versus Bonobo mtDNA (FST = 0.05–0.20); and Common Chimpanzee versus Bonobo (FST = 0.49–0.68). The implication is that this comparison between two anatomically modern human populations shows FST higher than Wright's threshold and higher than even some species differences! There are real problems with this comparison. First, the Salter FST calculation is much higher than virtually all other worldwide comparisons of FST or those between specific human populations. For example, Tischkoff and Kidd ("Biogeography of Human Populations") examined 369 SNPs and found FST between sub-Saharan Africans and Amerindians at 0.232. Given that the genetic distance between sub-Saharan Africans and Amerindians is greater than that between sub-Saharan Africans and Australoids, the Salter figures are hard to explain. In addition, the Tischkoff and Kidd study utilized SNPs that were from noncoding sections of DNA. Because this DNA does not produce protein or influence the regulation of genes that do, it is freer to accumulate genetic changes without facing negative NS (NS–). Thus, noncoding regions should produce much higher estimates of population divergence due to the absence of natural selection.

Garte actually calculated FST between Africans and Europeans by SNP category (nonsynonymous coding region, synonymous coding region, untranslated region, and intronic), utilizing data from the international HapMap project (S. Garte, "Human Population Genetic Diversity as a Function of SNP Type from HapMap Data," *American Journa'l of Human Biology* 22 [2010]: 297–300). In theory, FST should be highest for introns, untranslated regions, and synonymous coding regions, and lowest in nonsynonymous regions. This is exactly what they found (table 8.2). This result is consistent with Clark as well as Barreiro et al. and has profound significance for the question of racialized medicine, since genetic variation in the coding regions is going to be most significant for understanding the etiology of disease and drug response within and between populations (Clark et al., "Linkage Disequilibrium and Inference"; B. Barreiro, G. Laval, H. Quach, et al., "Natural Selection Has Driven Population Differentiation in Modern Humans," *Nature Genetics* 40 [2008]: 340–45). For example, Myles et al. examined worldwide population differentiation at disease-associated SNPs. They looked at FST from 25 SNPs associated with complex diseases (Crohn's, type 1 diabetes, type 2 diabetes, rheumatoid arthritis, coronary artery disease, and obesity). These were compared to the FST from 2750 randomly chosen SNP typed in the same individuals (N = 1000). The mean FST for the 2750 SNP's was about 0.100 with a variance of 0.004 (calculated from figure 3 in the Myles paper), and this was not statistically different from disease-associated SNPs (Myles et al. 2008). However, this study found that there are some high-risk alleles that are at very high frequency in some populations but at very low frequency in others. However, this fact doesn't always coincide with risk differences in a way that would be useful to the racial medicine theorists. For example, rs564398 is a SNP associated with type 2 diabetes. It is at very low frequency among the Kalash of Pakistan (the Central Asian mean frequency is 0.753) and in Melanesians (the Oceania mean frequency is 0.34). Yet, no one would consider these two groups "races" or even members of the same race.

TABLE 8.2 AFRICAN/EUROPEAN GENETIC DIVERSITY (FST) AS DETERMINED BY ALL CODING AND NON-CODING SNP'S WITHIN FIVE CHROMOSOMES

	FST Coding		FST Non-Coding		
Chrom No.	Mean +/- SEM	N	Mean +/- SEM	N	P
5	0.085 +/- 0.004	1,404	0.121+/- 0.001	83,283	< 0.0001
9	0.087+/- 0.005	1,149	0.122+/- 0.001	65,627	< 0.0001
13	0.074+/- 0.006	615	0.108+/- 0.001	44,692	< 0.0001
17	0.103+/- 0.005	1,403	0.154+/- 0.002	23,732	< 0.0001
21	0.098+/- 0.010	340	0.131+/- 0.001	17,787	< 0.0001

Source: S. Garte, "Human Population Genetic Diversity as a Function of SNP Type from HapMap Data," *American Journal of Human Biology* 22 (2010): 297–300

Finally, there is no direct comparison between FST as calculated by autosomal genes and mtDNA. The values for mtDNA will depend on whether the species in question is patrilocal (males stay put) or matrilocal (females stay put). These values can vary widely, for example, in plants mtDNA estimates can vary from biparental estimates by as much as 159 times (Hedrick, *Genetics of Populations*, 498). Primates vary on whether they are martilocal or partilocal and human cultures vary in this regard as well (H. Fishbein, and N. Dess, "An Evolutionary Perspective on Intercultural Conflict," in *Evolutionary Psychology and Violence: A Primer for Policymakers and Public Policy Advocates* [Westport, CT: Praeger, 2003]). Thus, there is really no way to interpret the FST data that Woodley presents unless you know exactly the mating structure of the species in question. For his extinct species comparisons, this cannot be done.

Thus, on balance, the notion that humans have biological races as measured by the criterion of genetic variation within and between population fails. However, this really is not the significant question from the point of view of evolutionary medicine. Evolutionary medicine posits that genes may be one source of disease. Evolutionary biology tells us that populations differ in gene frequencies, but it does not tell us that our socially constructed races correspond to the genetic variation that occurs in the human species. This is where the evolutionary and racialist medicine paradigms differ.

31. Templeton, "The Genetic and Evolutionary Significance of Human Races"; R. O. Andreasen, "A New Perspective on the Race Debate"; R. O. Andreasen, "The Cladistic Race Concept."
32. Templeton, "The Genetic and Evolutionary Significance of Human Races."
33. Ibid.; A. R. Templeton, "Genetics and Recent Human Evolution," *Evolution* 61, no. 7 (2007): 1507–19.
34. Linz et al., "An African Origin"; Handley, "Going the Distance."
35. J. L. Graves and M. R. Rose, "Against Racial Medicine," *Patterns of Prejudice* 40, nos. 4–5 (2007): 481–93.
36. M. I. Kamboh R. E. Ferrell, "Ethnic Variation in Vitamin D-binding Protein (GC): a Review of Isoelectric Focusing Studies in Human Populations," *Human Genetics* 72, no. 4 (1986): 281–93.
37. G. R. Sutherland, E. Baker, D. F. Callen et al., "Interleukin 4 is at 5q31 and Interleukin 6 Is at 7p15," *Human Genetics* 79 (1988): 335–37.
38. A. C. A. Clements, A. Garba, M. Sacko et al., "Mapping the Probability of Schistosomiasis and Associated Uncertainty in West Africa," *Emerging Infectious Diseases* 14, no. 10 (2008): 1629–32.
39. Tishkoff et al., "The Genetic Structure."
40. K. Bryc, A. Auton, M. R. Nelson et al., "Genome-wide Patterns of Population Structure."
41. J. L. Graves, *The Emperor's New Clothes: Biological Theories of Race at the Millennium* (New Brunswick, NJ: Rutgers University Press, 2005).
42. A. Helgadottir, A. Manolescu, A. Helgason et al., "A Variant of the Gene Encoding Leukotriene A4 Hydrolase Confers Ethnicity-specific Risk of Myocardial Infarction," *Nature Genetics* 38 (2005): 68–74, DOI:10.1038/ng1692.
43. G. C. Williams, "Pleiotropy, Natural Selection, and the Evolution of Senescence," *Evolution* 11, no. 4 (1957): 398–411; AP is described in P. B. Medawar, *An Unsolved Problem in Biology* (London: H. K. Lewis, 1952); M. R. Rose, *The Evolutionary Biology of Aging* (New York: Oxford University Press, 1991); L. Graves, "General Theories of Aging: Unification and Synthesis," in *Principles of Neural Aging*, ed. S. F. Dani, A. Hori, and G. F. Walter (Amsterdam: Elsevier Press, 1997), 35–55.
44. Graves, "Biological V. Social Definitions of Race."

45. M. Chang, M. L. Lindegren, M. A. Butler et al. "Prevalence in the United States of the Selected Candidate Gene Variants: Third National Health and Nutrition Examination Study, 1991–94," *American Journal of Epidemology* 169 (2009): 54–66.
46. K. E. Lohmueller, A. R. Indap, S. Schmidt et al., "Proportionally More Deleterious Genetic Variation in European Than in African Populations," *Nature* 451 (2008): 994–98.
47. J. L. Graves and M. R. Rose, "Against Racial Medicine," *Patterns of Prejudice* 40, no. 4–5 (2006): 481–493.
48. B. Charlesworth, *Evolution in Age-Structured Populations* (Cambridge, MA: Cambridge University Press, 1980).
49. W. D. Hamilton, "The Moulding of Senescence by Natural Selection," *Journal of Theoretical Biology* 12 (1966): 12–45; Charlesworth, *Evolution in Age-Structured*; Rose, *The Evolutionary Biology of Aging*; Graves, "General Theories of Aging: Unification and Synthesis"; M. R. Rose, H. Passanti, and M. Matos, *Methuselah Flies: A Case Study of the Evolution of Aging* (New Jersey: World Scientific, 2004).
50. Medawar, *An Unsolved Problem in Biology*, and Williams, "Pleiotropy, Natural Selection, and the Evolution of Senescence"; M. R. Rose and B. Charlesworth, "Genetics of Life History in *Drosophila melanogaster*: Exploratory Experiments," *Genetics* 97 (1981): 173–86; M. R. Rose, Laboratory Evolution of Postponed Senescence in *Drosophila melanogaster*," *Evolution* 38 (1984): 1004–10; Rose, *The Evolutionary Biology of Aging*.
51. G. B. Marin, M. H. Tavella, J. F. Guerreiro et al., "Short Communication: Absence of the E2 Allele of Apolipoprotein in Amerindians," *Brazilian Journal of Genetics* 20, no. 4 (1997), www.scielo.br/scielo.php?script=sci_arttext&pid=S0100-84551997000400029;H.C.Liu, C. J. Hong, S. J. Wang et al., "ApoE Genotype in Relation to AD and Cholesterol: A Study of 2,326 Chinese Adults," *Neurology* 53 (1999): 962; J. N. Henderson, R. Crook, J. Crook et al., "Apolipoprotein E4 and Tau Allele Frequencies in Choctaw Indians," *Neuroscience Letters*, 324, no. 1 (2002): 77–79; F. Willis, N. Graff-Radford, M. Pinto et al., "Apolipoprotein Epsilon4 Allele Frequency in Young Africans of Ugandan Descent Versus African Americans," *Journal of the National Medical Association* 95, no. 1 (2003): 71–76; W. M. Van der Flier, Y. A. Pijnenburg, S. N. Schoonenboom et al., "Distribution of APOE Genotypes in a Memory Clinic Cohort," *Dementia and Geriatric Cognitive Disorders* 25, no. 5 (2008): 433–38.
52. Marin et al., "Short Communication."
53. J. L. Graves, "Evolutionary Biology and Human Variation: Biological Determinism and the Mythology of Race," *Race Relations Abstracts* 18, no. 3 (1993): 4–34.
54. Graves, *The Race Myth*; X. Wu, V. Chen, B. Ruiz et al., "Incidence of Esophageal and Gastric Carcinomas Among American Asians/Pacific Islanders, Whites, and Blacks: Subsite and Histology Differences," *Cancer* 106, no. 3 (2005): 683–92.
55. I. Wu, D. Wu, F. Yu et al., "Association Between *Heliobacter pylori* Seropositivity and Digestive Tract Cancers," *World J. Gasteroenterology* 15, no. 43 (2009): 5465–71.
56. A. M. Bowcock, J. R. Kidd, J. L. Mountain et al., "Drift, Admixture, and Selection in Human Evolution: A Study with DNA Polymorphisms," *Proceedings of the National Academy of Sciences* 88 (1991): 839–43; S. A. Tishkoff, A. J. Pakstis, M. Stoneking et al.,

"Short tandem-repeat Polymorphism/Alu Haplotype Variation at the PLAT Locus: Implications for Modern Human Origins," *American Journal of Human Genetics* 67 (2000): 901–25; J. M Akey, G. Zhang, K. Zhang et al., "Interrogating a High-density SNP Map for Signatures of Natural Selection," *Genome Research* 12 (2002): 1805–14; S. A. Tishkoff and K. K. Kidd, "Biogeography of Human Populations: Implications for Race and Medicine," *Nature Genetics*, Suppl 1 (2004): S21–27; J. Z. Li, D. M. Absher, H. Tang et al., "Worldwide Human Relationships Inferred from Genome-wide Patterns of Variation," *Science* 319 (2008): 1100–04; Tishkoff et al., "The Genetic Structure."

57. Graves, *The Race Myth.*
58. Chang et al., "Prevalence in the United States."
59. L. Giot, J. S. Bader, C. Brouwer et al., "A Protein Interaction Map of *Drosophila melanogaster*," *Science* 302 (2003): 1727–36.
60. D. B. Goldstein, "Common Genetic Variation and Human Traits," *New England Journal of Medicine* 360, no. 17 (2009): 1696–98.
61. M. N. Weedon and T. M. Frayling, "Reaching New Heights: Insights Into the Genetics of Human Stature," *Trends in Genetics* 24, no. 12 (2008): 595–603; Goldstein, "Common Genetic Variation."
62. J. R. Behrman, R. Sickles, P. Taubman et al., "Black-White Mortality Inequalities," *Journal of Economics* 50, no. 1–2 (1991): 183–203; Graves, *The Race Myth.*
63. C. E. Finch and R. E. Tanzi, "Genetics of Aging," *Science* 278 (1997): 407–11.
64. Ibid.
65. J. v. B. Helmborg, I. Iachine, A. Skytthe et al., "Genetic Influence on Human Lifespan and Longevity, *Human Genetics* 119 (2006): 312–21.
66. J. Lee, A. Flaquer, R. Costa et al., "Genetic Influences on Life Span and Survival Among Elderly African Americans, Caribbean Hispanics, and Caucasians," *American Journal of Medical Genetics* 128A (2004): 159–64.
67. LaVeist, *Minority Populations and Health*, 24.
68. A. Desmond and J. Moore, *Darwin's Sacred Cause: How a Hatred of Slavery Shaped Darwin's Views on Human Evolution* (New York: Houghton, Mifflin, Harcourt, 2009).
69. R. Boyd and J. B. Silk, *How Humans Evolved*, 3rd ed. (New York: Norton, 2003).
70. Graves and Rose, "Against Racial Medicine," 481–493.
71. M. Gover, "Trends in Mortality Among Southern Negroes Since 1920," *Journal of Negro Education* 6 (1937): 276–88.
72. Graves, *The Myth of Race.*
73. H. Rose and P. D. McClain, *Race, Place, and Risk: Black Homicide in Urban America* (Albany, NY: SUNY Press, 1990); P. Dray, *At the Hands of Persons Unknown: The Lynching of Black America* (New York: Random House, 2002).
74. This is estimated from data presented in A. R. Permutt, J. Wasson, and N. Cox, "Genetic Epidemiology of Diabetes," *The Journal of Clinical Investigation* 115, no. 6 (2005): 1431–39.
75. M. Kouznetsova, X. Huang, J. Ma et al., "Increased Rate of Hospitalization for Diabetes and Residential Proximity of Hazardous Waste Sites," *Environmental Health Perspectives* 115, no. 1 (2007): 75–79.

76. National Center for Health Statistics 2003, www.cdc.gov/diabetes/.
77. Kouznetsova, "Increased Rate of Hospitalization."
78. R. D. Bullard, P. Mohai, R. Saha et al., *Toxic Wastes and Race at Twenty, 1987–2007* (United Churches of Christ Justice and Witness Ministries, 2007).
79. National Institutes of Health, Report 2011, "Estimates of Funding for Various Research, Condition, and Disease Categories (RCDC)," http://report.nih.gov/rcdc/categories/.

PART V

INTELLIGENCE AND RACE

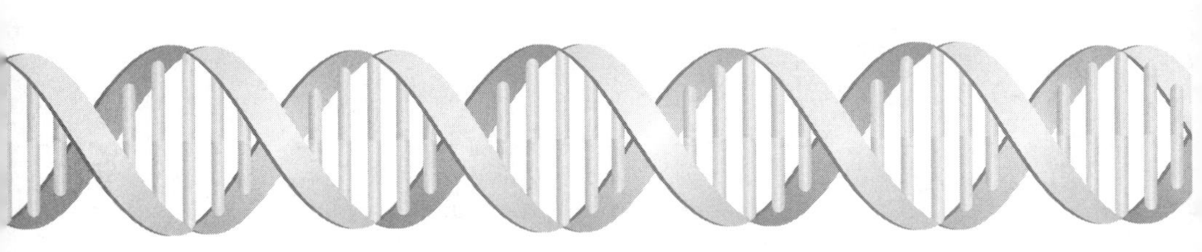

9

MYTH AND MYSTIFICATION

THE SCIENCE OF RACE AND IQ

Pilar N. Ossorio

Over the past two centuries biomedical science has, at times, provided justification for white privilege. Science has been used to support the proposition that differences in achievement reflect innate differences in ability among racial groups. Broadly speaking, the view that differences in academic achievement, IQ scores, employment status, or wealth primarily reflect innate differences is called "biological determinism."[1] As the late Stephen J. Gould pointed out, at its core, biological determinism is "a theory of limits. It takes the current status of groups as a measure of where they should and must be (even while it allows some rare individuals to rise as a consequence of their fortunate biology)."[2]

Biological determinism lost most of its scientific credibility by the mid-twentieth century, and lost much of its social and political power after World War II; however, it never entirely disappeared. Today, some people believe that persistent racial gaps in, for instance, school achievement, family income, and wealth must reflect innate differences in ability. Intelligence is one human trait postulated to play a role in many kinds of achievement, and some commentators hypothesize that racial differences in average levels of intelligence explain achievement gaps.

At the same time, the new molecular genetics has captured the public imagination and has provided tools for conducting large-scale genetic comparisons between individuals and between human groups. Some people will look to modern genetics to provide scientific justifications for racial inequalities. Genetics is particularly appealing in this role because of its apparent precision, authority, and high-tech chic. Many people reason that if groups vary with respect to innate cognitive abilities, then the differences between groups must be attributable to differing racial patterns of genetic variation; they suppose that the races possess different "intelligence genes." To disentangle claims about race, genetics, and intelligence, we must examine beliefs about race and intelligence, and understand what role genes reasonably could or could not play with regard to the intersection of these two concepts.

RACE AND GENETICS

Race is a concept that people use to make sense of the amazing panoply of human biological, social, cultural, and political variation. Since the seventeenth century, Western scientists have attempted to categorize and differentiate human populations. Perhaps the single most influential taxonomy of human races was that constructed by the Swedish naturalist Carolus Linnaeus, in his famous *Systema Naturae*. The 1758 edition of this work initiated the modern science of systematically categorizing all living organisms. Linnaeus included humans in the *Systema* as an ordinary component of the natural world, subject to categorization just like any other organism. He identified four geographically based human races, which he designated as subspecies. These four groupings are described in contemporary terms as Native Americans, Europeans, Asians, and Africans.

Linnaeus's taxonomy described both the physical appearance and personality type that he believed characterized each race. *Homo sapiens americanus* (Native Americans) were red in color, had erect posture, straight and thick black hair, wide nostrils, scanty beards, and "harsh" faces. They were, purportedly, liberty-loving, ill-tempered, and obstinate, and their social and political relations were governed by custom rather than formal law. *Homo sapiens europeaeus* (Europeans) had white skin, long flowing hair, and blue eyes. Their character was serious, sanguine, and smart, and their social and political relations were governed by formal law. *Homo sapiens asiaticus* (Asians) had yellow skin, black hair, and dark eyes. According to Linnaeus, they were melancholy, greedy, and governed by "opinion" rather than formal law. *Homo sapiens afer* (Africans) were black, with "frizzled" black hair, silky skin, flat noses, and "tumid lips." He described their character as crafty, impassive, lazy, careless, and "ruled by caprice."[3]

Obviously, Linnaeus was not simply cataloging physical features or biological structures, he was also using nonbiological traits (real or imagined), such as dress and specific forms of social organization, to construct his categories. He commingled sociocultural traits and biological ones as though all had the same cause. Linnaeus may have believed that he was simply performing "objective science," analyzing human beings the same way that he would analyze snakes or anteaters, but he was not. Linnaeus was constructing a social hierarchy in which whites (the group to which he belonged) were at the pinnacle and blacks were at the bottom. Whites possessed the qualities he viewed as most admirable and valuable, blacks possessed the least of those qualities, and other races were somewhere in between.

From Linnaeus's day until the twentieth century, the overarching goal of the study of human variation was to establish a small number of basic categories into which all human variation could be fit. Anthropologists collected large quantities of data, such as body measurements of people from around the world, while attempting to discover discrete, natural categories of people. But the more people the anthropologists measured, the more categories they found. Linnaeus's four geographic categories are still evident, however, in contemporary "folk notions" of race. The link between the continent of ancestry, a person's appearance, and her race is still central in most people's belief systems.

By the early twentieth century, many scientists were beginning to believe that nature had not produced discrete, biological categories of humans. One reason scientists had difficulty allocating the world's people into categories is because the different physical traits that scientists measured, including pigmentation, hair texture, facial features, leg lengths, etc., do not vary in concert with each other. Consider two features that many people use as bases for racial categorization—skin color and nose shape. Dark skin color and wide nose shape are features associated with people from sub-Saharan Africa, and with a race designated as black. Light skin color and narrow, longer noses are features associated with people from Europe, and with a race variously designated as white,

Caucasian, or Anglo. But some individuals have dark skin color and narrow, long noses. This is true of many people from the Middle East, South Asia, and Northern Africa. Skin color and nose shape are not always correlated, they vary independently of each other, and so using these two features can lead to more than two ways of categorizing people.

Eventually, the accumulated anthropological data showed that patterns of human biological variation are complex, and that biological variation between human groups is generally a matter of gradual change rather than abrupt discontinuities. On average, people tend to be more physically similar to others who are born and live near them, and one group gradually gives way to another. Charles Darwin argued that the weightiest of all arguments against the notion of distinct human races was that "they graduate into each other . . . [the naturalist] of a cautious disposition . . . will say to himself that he has no right to give names to objects he cannot define."[4] Because human biological variation is complex and continuous, allocating people to categories requires us to "draw lines" where none exist in nature.

Differences in features and, as it turns out, in DNA sequences are greatest between groups of people who are geographically distant from each other. The pattern in which some measurable feature varies gradually, and that variation correlates with geographic distance, is called "clinal" variation.[5] Many human traits and many human genetic differences exhibit a clinal pattern.[6]

Modern molecular genetics has largely confirmed what anthropology discovered in the early twentieth century. As a general rule, all people possess the same genes, in the same order, on their chromosomes. Genomes (the complete complement of DNA in a cell) of people from all points on the globe are remarkably similar. This is why it was possible and reasonable to undertake the Human Genome Project (HGP) to create a composite reference DNA sequence that was pieced together from bits of several people's genomes.[7]

Although humans are quite genetically similar, no two people are genetically identical unless they are monozygotic (identical) twins. There are various methods of conducting genetic comparisons between individuals or groups, and the different methods do not always give exactly the same results, so claims about genetic similarity and difference should be made cautiously, and numerical claims should be viewed as estimates or approximations. Given this caveat, contemporary geneticists generally accept that any two unrelated humans are about 99.5 percent to 99.9 percent genetically identical.[8] But because the human genome contains approximately 3 billion nucleotides (DNA building blocks), a 0.1 percent or 0.5 percent difference translates into millions of sites at which two people will have a different nucleotide. Some genetic variation involves differences in the number of copies of a particular DNA sequence a person possesses (copy number variation). Copy number variation has been linked to human differences in drug response, resistance to HIV infection, risk of developing autism or schizophrenia, and other human traits.[9]

Nearly all of the genetic variation among humans is found between individuals within any human group. If one assesses the genetic variation in a group of Yoruba people from Nigeria, and in a group of Swedish people from the city of Malmo, somewhere between 85 percent and 95 percent of the genetic variants will be found in both groups, although some variants will be found at a higher frequency in one group than the other.[10] Furthermore, if one examines a single gene or region of the genome, then an individual whose recent ancestors are from Lagos, Nigeria, may be more similar to somebody from Malmo, Sweden, than to most other people from Nigeria.[11] Depending on how one measures, the component of genetic variation that occurs between human groups from different continents could be as low as 2.8 percent, whereas the component of genetic variation between human groups from the same continent could be 2.5 percent.[12] If one associates races with particular continents, then all but 2.8 to 5 percent of the human genetic variation is found within any race. Furthermore, very rare genetic variation may be associated with only one family or found in only one person, and will play no significant role in between-group comparisons.

When the frequency of a gene variant does differ between one human group and another, one typically sees a pattern where, for instance, gene variant A is found in 15 percent of people from group 1 and found in 23 percent of people from group 2. Typically, the difference in frequency is relatively small and most people will have the same variant, regardless of the group to which they belong. There are, however, a few gene variants that differ quite significantly from one human group to another.

The gene with highest degree of between-group variation is one that influences skin pigmentation in humans and other organisms—the

SLC24A5 gene.[13] One version of this gene has been found in 98 to 100 percent of people in European population samples studied, but a different version of this gene has been found in 93 to 100 percent of people sampled from parts of Africa, East Asia, and the Americas. Interestingly, over half of the African Americans studied have either one or two copies of the gene variant commonly found in European populations.

To the extent that genes underlie pigmentation, perhaps the most visible of human differences, we should expect that some of those genes will have quite different frequencies among people of different skin color. What may be surprising is that pigmentation is genetically very complex. At least five different genes strongly influence skin pigmentation, and there are hundreds of other genes that play a minor or occasional role.[14] Because of this complexity, scientists cannot use variants of pigmentation genes as a means of racial classification. There are no pigmentation gene variants that are found only and always in "white people" but not in people of other racial groups.

Human beings are quite genetically similar because humans are a relatively young species, one that has not had much time to differentiate (as measured by number of generations).[15] No human group has been reproductively isolated from others for long enough to become a different species or subspecies.

The greatest amount of genetic variation between human individuals is found among people of Africa. For many regions of the human genome, there are more variants found among people of Africa (and the recent African diaspora) than among people in the rest of the world. This is probably because humans have resided in Africa for much longer than we have resided any place else in the world, so our species has had time to accumulate genetic changes within the people in Africa. A relatively small group of people migrated out of Africa thirty to one hundred thousand years ago, and only a fraction of the human genetic variation went with them.[16] After migrating out of Africa, humans underwent a very rapid expansion to all parts of the globe. One implication of the high degree of genetic diversity among people of Africa is that it is incorrect and incoherent to think of black people as a genetically unitary group.

Scientists are intensely interested in using the 5 to 15 percent of genetic variation that occurs between populations to study the history of the human species. Understanding this variation also helps them to conduct studies on genetic causes of human disease. Recently, geneticists have begun measuring hundreds of thousands of sites of genetic variation in each of thousands of people's genomes, then statistically grouping people according to their overall patterns of genetic similarity and difference. Such analyses can group together people whose ancestors all came from the same continent, and differentiate them from people whose ancestors came from a different continent.[17] Some commentators have equated these statistical groupings with human races.

But the ability to statistically cluster members of the human species is not the same thing as "finding" a natural, biological category.[18] Using the same scientific approach, scientists can also statistically group people whose grandparents were born in Finland and separate them from people whose grandparents were born in Sweden.[19] Researchers can group people of Iceland according to the county or counties in which their ancestors (five generations back) were born.[20] This does not mean that the people of Finland and Sweden are of different races, or that Iceland's counties are populated by different races. Also, keep in mind that only about 0.1 to 0.5 percent of the human genome differs from one person to another, and only 5 to 15 percent of that variation can be used to distinguish between human groups, so statistical grouping of people by patterns of genetic variation is based on a minute fraction of our genomes.

Finally, it is crucial to reemphasize that the amount of genetic variation between groups is very small compared to the 85 to 95 percent of variation found between different individuals within human groups. That one can genetically group together people whose four grandparents all came from Beijing, China, and differentiate them from people whose four grandparents all came from someplace else does not mean that the people whose grandparents came from Beijing are substantially genetically alike. People in any particular place are not genetically homogenous, and people certainly are not homogenous across entire continents. The vast majority of human genetic variation is between individuals, including individuals who can be assigned to the same racial, ethnic, or national group.

Because humans have high within-group genetic variation, genes are unlikely to explain average differences in IQ test scores of different racial groups. We do not know the extent to which genes underlie a person's ability to perform complex mental tasks, but there is no reason to think that people whose relatively recent ancestors all came from one continent would have different variants of any relevant genes than people whose ancestors came from another continent. If potential "cognition genes" are similar to other genes, then most variants will be found within all groups of people at similar frequencies.[21]

WHAT IS RACE?

> "Order is Heaven's first law: and, this confessed,
> Some are, and must be, greater than the rest . . .
> Without this just gradation, could they be
> Subjected, these to those or all to thee?"
>
> —Alexander Pope, *Essay on Man* (1733)[22]

If races are not distinct genetic groups of human beings, then what are they? Some scholars have argued that because humans are so genetically similar, and because there are no discrete genetic groups, human races do not exist or are not "biologically relevant."[23] However, the issue of whether something exists, and whether it is "biologically relevant," is completely separable from the question of whether it can be found in our genes.

Many nongenetic features of our world and ourselves are real because we make them so, we bring them into existence through beliefs, customs, laws, physical arrangements of our environment, and numerous everyday acts. Marriages, schools, and subways are not encoded in any person's genes, but they are all real. Likewise, race is real because people believe in it and act on those beliefs. Race is deeply rooted in the consciousness of individuals and groups, and it structures our lives and our physical world in myriad ways. It is a strong predictor of where people live,[24] what schools they attend,[25] where and how their spirituality is practiced, what jobs they have, and the amount of income they will earn.[26] Race is real because human beings continually create and recreate it through the process of racialization.[27]

There is no unitary definition of race, no definition that applies in all places, at all times, and for all purposes. Scholars who include race as a variable in their studies must operationalize the concept of race in a manner that meets the needs of their study, while acknowledging that such "working definitions" merely "fulfill the need for an analytical strategy, they do not reflect a fixed social or biological reality."[28]

Scholars who study race generally agree on a few themes regarding how notions of race operate in society. One point of agreement is that race is a second-order construct—a belief about beliefs, behaviors, and traits. A person's racial self-identification, and her ascription of race to other people, is based on her beliefs about skin color, head shape, hair texture, religion, ancestry, language spoken, nationality, dress, political philosophy, and many other factors. Genes encode some of the traits on which people base racial attributions and identity, but genes do not encode all of them. Because a person's race is not reducible to her genetically encoded traits, one could know everything about a person's genome and still not know her race.[29]

Another point of agreement among race scholars is that race is a very malleable concept. This claim developed from numerous observations of how the meaning of racial categories, and the categories themselves, change over time.[30] One well-known example of this malleability is that every decennial census since the early twentieth century has defined race differently than the previous census.[31] Thus, a person who was white in one census might have been Mexican or black in another. Another example is that some people have a different race on their birth certificate and their death certificate.

The malleability of the concept of race means that one must be very careful in comparing different studies in which race is correlated with some other feature, such as income or IQ. Studies done at different times, or by different researchers, may not be measuring the same people or the same social reality when they refer to, for instance, Native Americans or white people. Furthermore, because race plays such a powerful role in shaping our lives, it is correlated with many factors that likely influence cognitive ability. These correlated factors may confound the results of studies by making it appear that race explains or "causes" between-group differences in IQ test results when some other racially correlated factor, such as diet, parental income, neighborhood environment, etc., actually explains the differences.

Another point of agreement among race scholars is that race involves practices that create hierarchies of social privilege and deprivation. While individuals may move from one racial group to another, the practices and social institutions that create racial hierarchy are long-standing and deeply entrenched. Some group is always at the pinnacle of privilege, and in the United States this has been white people. This feature of race has not changed since Linnaeus's and Alexander Pope's time. The concept of race emerged in Western culture concomitant with European colonialism; race was one ideology that could be used to justify conquest.[32] In the Americas, the asserted racial superiority of white or European-descended people functioned to justify the enslavement of African-descended people, the near extermination of Native Americans, and the extremely oppressive labor policies directed toward Chinese and Japanese workers. Contemporary racial hierarchies are enforced by less explicit means, but they still operate to rank people on scales of social status and privilege.

In the contemporary world, beliefs about racial difference and racial superiority or inferiority may be articulated in the language of molecular genetics and genomics. Modern genetics has great authority, and beliefs about race that once relied on vague notions of innate difference can be made to sound more precise and credible by framing them as genetic explanations. A person would then look for genetic differences between racial groups for the explanation of racial differences in achievement. Educational achievement, wealth, and other measures of status often run in families, a fact that may increase the intuitive credibility of genetic explanations. Societal institutions operate to entrench groups who wield power into self-perpetuating dynasties. From the Tudor monarchical dynasty in sixteenth-century England to the Bush and Kennedy family dynasties in the twentieth and twenty-first centuries in America, families with access to power pass their positions of privilege on to succeeding generations in a process resembling the hereditary transmission of genetic traits.

RACE AND INTELLIGENCE

Just as there is no unitary definition of race, there is no agreed-upon or single definition of intelligence. One aphorism holds that intelligence is what intelligence tests measure. Psychometricians argue that intelligence tests measure reasoning skills, although the tests also measure knowledge.[33] Some innovative scholars have developed theories of emotional intelligence and multiple intelligences—multiple types of cognitive function that are valuable and measurable, and that may manifest differently in different contexts.[34] However, the typical IQ test does not measure multiple intelligences; instead, the test produces a single intelligence quotient.

Some scholars argue that one's IQ indicates one's general cognitive ability, or *g*.[35] Many other scholars argue that the notion of a single, general quality that underlies performance on all cognitive tests is incoherent.[36] Stephen Jay Gould has provided a thorough explanation and critique of the concept of *g* in *The Mismeasure of Man*.[37] *g* has been a useful for commentators who seek to create social hierarchies based on intelligence, because "ranking requires a criterion for assigning all individuals to their proper status in a single series."[38]

Alfred Binet, the developer of the first intelligence test in the early twentieth century, rejected the notion that his test measured a person's inborn or fixed cognitive ability. He also declined to use his test to rank individuals according to cognitive ability. The purpose for which he devised the test, and the only purpose for which he thought it appropriate, was to measure the intellectual capacity of children who were performing poorly in school, to determine which children had cognitive deficits for which remedial instruction might be helpful. Later psychologists, particularly those in the United States, took up and modified Binet's test and were willing to embrace the view that intelligence was an inborn and fixed attribute of a person. We can call this view the hereditarian theory of IQ.

Over the past decade, some contemporary proponents of the hereditarian theory have argued that (1) IQ is the most important determinant of academic success, (2) academic success is the most important determinant of high status and wealth-generating employment, and therefore (3) the economic elite have their positions and wealth as a matter of merit (intellectual contribution to society), and conversely, members of the economic underclass also deserve their position at the bottom of the social hierarchy.[39] These commentators argue that programs aimed at raising the academic achievement of disadvantaged students are misguided because those students are, on average, genetically incapable of significant academic success. Hereditarians view racial gaps in test scores, from IQ to the SAT, as evidence of inherent and immutable racial or ethnic differences in underlying cognitive capacity.

Many claims of contemporary hereditarians have been critiqued and debunked in books such as *Measured Lies*, *Inequality by Design*,

Whitewashing Race, and *Intelligence and How to Get It*.[40] These books describe mistakes of fact, method, and logic made by the hereditarians. A significant problem in debates about hereditarian theories of IQ is that correlation is often treated as proof of causation. If one observes that people in lower socioeconomic brackets, on average, score lower on IQ tests than people in higher socioeconomic brackets, this does not mean that low IQ causes poverty. It could be that poverty causes low IQ,[41] or that something else causes both outcomes. If IQ test scores correlate with race (however race is defined), this does not mean that some inborn racial essence causes particular IQ test scores. One reasonable alternative explanation is that race is correlated with other factors, such as quality of schools, exposure to lead, or malnutrition, and these other factors are causing the observed differences in test scores.

Hereditarian claims are based on the alleged heritability of IQ. Heritability assesses the way a trait *varies in a population* and purports to measure how much of that variation is explained by genetic differences *within* the population. The remaining variation is attributed to all other factors (the environment and nongenetic aspects of biology). If children in a classroom score between 90 and 130 on an IQ test, a hereditarian might claim that 65 percent of the 40 point difference in IQ is due to genetic differences between the students, and 35 percent is due to other factors. Strong proponents of hereditarian theories tend to believe that genetic differences explain as much as 85 percent of the variation in adult IQ in a population, but other scholars believe that genes explain much less than 50 percent of the variation in IQ.

Heritability is a population measure and it is only valid within the group in which the trait of interest was studied. Thus, if one measured IQ scores in a group of middle-class white children and a group of middle-class Native American children, and found that the average IQ score for the Native American children was 3 points higher than for the white children, one could specify a heritability score for each group, *but one could not say how much of the difference between the white and Native American children's mean score was attributable to genes.*

To illustrate why heritability within groups has nothing to do with heritability between groups, consider a hypothetical example in which researchers measure the height of men sampled randomly across the state of Wisconsin and men sampled randomly across the island of Fiji.

Researchers might determine that the differences in height in Wisconsin men were primarily due to genes, and that the same was true in Fiji. However, if men in Wisconsin were an average of two inches taller than the men in Fiji, researchers could not state that the two-inch between-group difference was caused by genetics. Different diets might explain most or all of the between-group difference. This example should illustrate why, *by itself, a difference in the mean IQ scores between two groups of people can never justify the inference that some or all of that difference is caused by genetics.* Even scholars with a strong hereditarian bent acknowledge this point. To the extent that children from different racial groups experience systematically different types of education, peer groups, toxic exposures, medical care, diets, etc., they should be viewed as belonging to different "groups" for purposes of assessing the heritability of IQ.

Many scholars question the entire enterprise of treating heritability statistics as though genes and environment are actually separable influences on IQ or any other trait. Genes always function within particular environments to shape the developing human organism. The developmental interaction among many genes, and numerous environmental factors, is complex, varies over time, and is susceptible to chance events.

Researchers have found evidence that some environmental factors are strongly associated with IQ and other measures of cognition. Malnutrition and exposure to environmental toxins, such as lead from paint, are strongly correlated with IQ. The quality of a person's school significantly impacts her IQ score—children who begin their education in poor quality schools and then move to better ones show increases in their IQ scores.[42] A study published in 2009 found that long-term stress is negatively associated with young adults' performance on cognitive tests.[43] This study measured levels of several physiological properties associated with stress, including blood pressure, cortisol, and epinephrine levels. The researchers collected data throughout their participants' childhood years, then administered tests of cognitive performance when the children turned seventeen years old. Young adults whose bodies exhibited the highest levels of chronic stress had the least effective working memories and poorer cognitive performance. Correlation does not prove a causal relationship between environmental variables and IQ any more than it proves a causal relationship between race and IQ, but the presence of many strong environmental correlates with IQ raises doubts about racial or genetic explanations.

Research also undermines the hereditarian claim that IQ is the primary determinant of achievement. Many environmental variables predict achievement as well as or better than IQ,[44] except for people whose IQ scores are at the abnormally low end of the scale. For instance, a person's social environment may be an important determinant of her achievement, yet variables that capture a person's social environment are often, literally, left out of the equation in work done by hereditarians. The social environment includes the expectations of one's peers, encouragement by one's parents and teachers, enrichment opportunities available in the neighborhood, etc. A decades-long study that included social environment variables found that a 15 point difference in IQ scores among high school boys only explained 6 percent of the variability in their earnings at age thirty-five. The greater the number of social factors taken into account, the less important IQ became.[45] Childhood social-context variables were still significantly correlated with earnings by age fifty-five. In a related analysis, Fisher et al. demonstrated that if all adults in the country had the same score on an IQ test, the variation in household income would only decrease by about 10 percent. Contrary to hereditarian claims, these data suggest that differences in IQ do not explain much about professional achievement and wage inequality, including wage inequality between racial groups.[46] On the other hand, factors external to an individual can greatly influence her or his lifelong course of achievement.

Because race comprehensively structures people's lives in the United States, it is correlated with many environmental factors that can influence IQ and achievement. People of different races tend to live in different neighborhoods, so they may be exposed to different levels of lead, different quality schools, different diets, and different levels or types of stress. They may be exposed to different attitudes about achievement. People of minority groups may routinely experience racism, a kind of stress that can have long-term physiological consequences. On average, people of different races receive health care at different institutions, and the care they receive is not of the same quality. In sum, racial groups differ with respect to so many environmental factors that it is entirely plausible that environmental differences explain current racial gaps in mean IQ scores.

The environment can be modified in ways that genes cannot. When the environment is changed, the trait of interest (in this case intelligence) may also change *even though genes also play a role in shaping that trait.* For instance, in one study African American children in Milwaukee who were thought to be at risk for cognitive disability were randomized so that half received intensive day care and early, enriched education, while the other half received ordinary day care and schooling.[47] By age five, children who received the intensive intervention averaged 110 on a standard IQ test (above average), while children in the control group averaged 83 (well below average). The effects of early, intensive education were still apparent by adolescence, when the children from the intervention group scored, on average, 10 points higher on IQ tests than the children from the control group.

There is some evidence that differing environments have influenced the entire human population's IQ scores over time. People's average IQ scores have risen by about 3 IQ points per decade over the last century.[48] The average IQ score from 1917 would amount to about 73 on today's tests. This effect almost certainly is not due to changes in human genetics, because there has not been enough time for new intelligence-related mutations to arise and spread throughout human populations. Nor has there been enough time for intelligence-related genetic variants already in the population to significantly increase or decrease in frequency. The most likely explanation for the rise in IQ is that some relevant environmental factors have changed, causing people to develop in ways that are reflected in higher average IQ scores.

Another piece of evidence concerning widespread environmental influences on IQ is that the mean difference between black Americans' and white Americans' test scores has narrowed since the 1970s. Using data from several different IQ tests that were administered in a standard manner to black and non-Hispanic white people, Dickens and Flynn showed that blacks have narrowed the IQ gap by one-third to one-half of what it was in the 1970s.[49] If IQ were a fixed, intrinsic quality of races, then the IQ gap should be stable over time, but it is not.

GENES, BRAINS, AND INTELLIGENCE

It is quite difficult to study the intersection of genes, brains, and cognition. Presently, scientists have only vague and preliminary ideas about how brain structures correlate with thought processes (including solving problems on intelligence tests), and scientists are only beginning to study the ways in which genes influence the development of brain structures. One can imagine various aspects of brain physiology that might influence thinking. These include the speed and efficiency of communication between brain cells or structures, the quantity of cells that can be mobilized to store a memory or solve a problem, the density of receptors for neurotransmitters, etc. To date, none of these aspects of physiology has been demonstrated to play a role in normal human problem solving or abstract thinking ability, but it seems reasonable that many aspects of brain physiology could play a role. Hundreds or thousands of genes, operating in particular environmental contexts, encode information that is important for the development and maintenance of brain structures and physiological processes.

Some researchers have attempted to correlate genetic markers with people's scores on tests of cognitive ability. Thus far, these studies have yielded some claims about gene variants that are correlated with variation in cognitive ability, but no research has demonstrated a *causal* connection between a particular gene variant and a particular degree of cognitive skill (within the normal ability ranges). Even proponents of such research note that "there are an unknown number of genetic influences on different abilities; [and] some . . . proportion of these may be too small to feasibly be detected."[50] When variants of particular genes have been associated with IQ, the effects reported are relatively small—in the range of a couple of IQ points—and few of the observed correlations have been replicated. In contrast, altering environmental factors, such as the quality of education and day care, has been associated with IQ variation of 10 points or more.[51]

In 2005, a group of scientists created a firestorm of controversy when they reported evidence for recent natural selection in humans of variants in two genes involved in regulating brain size.[52] These scientists reported that a variant of the *MCHP1* gene arose approximately thirty-seven thousand years ago and rapidly became prevalent in all populations tested except those in Africa.[53] Their companion paper reported that a variant of the *ASPM* gene arose much more recently, approximately five to ten thousand years ago, and became prevalent in Europe and the Middle East.[54] The authors noted that the spread of the *ASPM* variant occurred at around the time when humans domesticated crops and livestock, when urbanization occurred and written languages developed. The authors all but invited readers to infer that a particular gene variant, which had been selected for in white people but not in other racial groups, was importantly involved in the development of civilization.

A flurry of media discussion and controversy surrounded the "brain gene" papers. The claim that gene variants for head size had been positively selected in people with recent ancestors from certain continents but not others was particularly notable because of the long-standing, but highly contested, claim that head or brain size is correlated with IQ. Attempts to correlate head size, IQ, and race or gender go back to the nineteenth century, during which time European male scientists asserted that women and nonwhites (particularly people from Africa and the Americas) were intellectually inferior to European men, and that this inferiority could be "objectively" demonstrated by measuring head sizes.[55] The science of earlier centuries has been refuted, but many people saw the 2005 papers on *MCHP1* and *ASPM* as reintroducing those old arguments dressed up in the garb of modern molecular science.

Since 2005, other researchers have evaluated the same data on *MCHP1* and *ASPM*, plus some additional data, and concluded that there is no evidence that these genes have been under natural selection in modern humans.[56] These reanalyses undercut the idea that the particular variants found at high frequency among people of European descent somehow made Europeans better adapted for modern civilization. Additional studies have discovered that the *MCHP1* and *ASPM* variants reported in the 2005 papers do not correlate with larger (or smaller) than average head size.[57] The genes were originally described as having to do with head size because some variants of these genes can cause microcephaly (extremely small heads that lack major portions of the brain). However, those microcephaly-causing variants were not included in the studies published in 2005. Finally, several research groups have tried *and failed* to show any correlation between the variants described in the 2005 papers and IQ, reading abilities, or verbal abilities.[58] One article stated, "The results strongly did not support the hypothesis that [variants] in *MCPH*-related genes are related to the evolution of human language or cognition."[59] Unfortunately, none of these follow-up studies received media attention.

CONCLUSION

The binary formulation of "genes versus environment" is misleading. Cognitive abilities are complex and will likely be influenced by a myriad of environmental factors *and* genes. Given the complexity of brains and cognition, one ought not expect that a few genes will play a dominant role in shaping the normal range of human cognitive abilities; numerous genes will be involved. It is statistically implausible that variants of numerous genes relating to intelligence would be distributed among racial groups in a manner that systematically confers cognitive advantage on one group or disadvantage on another. Furthermore, there is no evidence to support the claim that current racial differences in mean IQ scores are caused by racially distinctive patterns of genetic variation.

There *is* evidence that IQ scores are influenced by environmental factors that are pervasively and systematically patterned along racial lines in the United States. Nonetheless, mean IQ differences among racial groups have been decreasing over the past few decades, perhaps in response to improved educational opportunities for some minority individuals (e.g., the affirmative action programs began in the 1970s). Taken together, the evidence suggests that differences in IQ scores are the *result* of social inequality rather than its cause.

NOTES

1. Anonymous, "Intelligence and Genetic Determinism," *GeneWatch* 19 (2006): 9–12.
2. S. J. Gould, *The Mismeasure of Man* (New York: Norton, 1981), 28.
3. J. Marks, *Human Biodiversity: Genes, Race, and History* (New York: Aldine De Gruyter, 1995).
4. Charles Darwin, *The Descent of Man, and Selection in Relation to Sex* (Princeton, NJ: Princeton University Press, 1981), 698 (originally published in 1871).
5. L. J. L. Handley, A. Manica, J. Goudet et al., "Going the Distance: Human Population Genetics in a Clinal World," *Trends in Genetics* 23 (2008): 432–39.
6. D. Serre and S. Paabo, "Evidence for Gradients of Human Genetic Diversity Within and Among Continents," *Genome Research* 14 (2004): 1679–85.
7. International Human Genome Sequencing Consortium, "Finishing the Euchromatic Sequence of the Human Genome," *Nature* 431 (2003): 931–45; International Human Genome Sequencing Consortium, "Initial Sequencing and Analysis of the

Human Genome," *Nature* 409 (2001): 860–921; J. C. Venter, Mark D. Adams, Eugene W. Myers et al., "The Sequence of the Human Genome," *Science* 291 (2001): 1304–51.

8. A. Chakravarti, "To a Future of Genetic Medicine," *Nature* 409 (2001): 822–23; S. A. Tishkoff and K. K. Kidd, "Implications of Biogeography of Human Populations for 'Race' and Medicine" *Nature Genetics* 36 (2004): S21–S27; L. B. Jorde and S. P. Wooding, "Genetic Variation, Classification and 'Race,'" *Nature Genetics Supplement* 36 (2004): S28–33.
9. J. A. Buchanan and S. W. Scherer, "Contemplating Effects of Genomic Structural Variation," *Genetic Medicine* 10 (2008): 639–47; E. H. Cook and S. W. Scherer, "Copy-Number Variations Associated with Neuropsychiatric Conditions," *Nature* 455 (2008): 919–23; I. Johansson and M. Singelman-Sundberg, "CNVs of Human Genes and Their Implications in Phamarcogenetics," *Cyogenetic and Genome Research* 123 (2008): 195–204; F. Speleman, C. Kumps, K. Buysse et al. "Copy Number Alterations and Copy Number Variation in Cancer: Close Encounters of the Bad Kind," *Cytogenetic and Genome Research* 123 (2008): 176–82.
10. L. L. Cavalli-Sforza and M. W. Feldman, "The Application of Molecular Genetic Approaches to the Study of Human Evolution," *Nature Genetics Supplement* 33 (2003): 266–75.
11. Jorde and Wooding, "Genetic Variation."
12. N. A. Rosenberg, J. K. Pritchard, J. L. Weber, and H. M. Cann, "Response to Comment on 'Genetic Structure of Human Populations,'" *Science* 300 (2003): 1877.
13. R. L. Lamason, M. P. K. Mohideen, J. R. Mest et al., "SLC24A5, a Putative Cation Exchanger, Affects Pigmentation in Zebrafish and Humans," *Science* 310 (2005): 1782–86.
14. Ibid.; G. S. Barsh, "What Controls Variation in Human Skin Color," *PLoS Biology* 1 (2003): 19–22.
15. Jorde and Wooding, "Genetic Variation"; H. C. Harpending, M. A. Batzer, M. Gurven et al., "Genetic Traces of Ancient Demography," *PNAS* 95 (1998): 1961–67.
16. Handley et al., "Going the Distance"; Tishkoff and Kidd, "Implications of Biogeography"; S. Olson, *Mapping Human History: Discovering the Past Through Our Genes* (New York: Houghton Mifflin, 2002); Tishkoff et al., "The Genetic Structure and History of Africans and African Americans," *Science* 324 (2009): 1035–44.
17. Tishkoff and Kidd, "Implications of Biogeography"; Jorde and Wooding, "Genetic Variation"; N. A. Rosenberg, J. K. Pritchard, J. L. Weber et al., "Genetic Structure of Human Populations," *Science* 298 (2002): 2381–85.
18. R. S. Cooper, "Race and IQ: Molecular Genetics as *Deus ex Machina*," *American Psychologist* 60 (2005): 71–76.
19. Diabetes Genetics Initiative et al., "Genome-Wide Association Analysis Identifies Loci for Type 2 Diabetes and Triglyceride Levels," *Science* 316 (2007): 1331.
20. A. Helgason, B. Yngvadottir, B. Hrafnkelsson et al., "An Icelandic Example of the Impact of Population Structure on Association Studies," *Nature Genetics* 37 (2005): 90–95.
21. Cooper, "Race and IQ: Molecular Genetics."

22. Gould, *Mismeausre*, 31.
23. S. B. Haga and J. C. Venter, "FDA Races in the Wrong Direction," *Science* 301 (2003): 466–67; R. S. Schwartz, "Racial Profiling in Medical Research," *New England Journal of Medicine* 344 (2001): 1392–93.
24. D. S. Massey, "Residential Segregation and Neighborhood Conditions in U.S. Metropolitan Areas," in *America Becoming: Racial Trends and Their Consequences*, vol. 1, eds. N. J. Smelser, W. J. Wilson, and F. Mitchell (Washington, DC: National Academies Press, 2001); D. S. Massey and N. Denton, *American Apartheid: Segregation and the Making of the Underclass* (Cambridge, MA: Harvard University Press, 1993.
25. G. Orfield and C. Lee, *Why Segregation Matters: Poverty and Educational Inequality* (Cambridge, MA: Harvard Civil Rights Project, 2005).
26. R. M. Blank, "An Overview of Trends in Social and Economic Well-being by Race," in *America Becoming*.
27. M. Omi and H. Winant, *Racial Formation in the United States*, 2nd Edition (New York: Routledge, 1994).
28. Smelser et al., *America Becoming*, 4.
29. P. Ossorio, "About Face: Forensic Genetic Testing for Race and Visible Traits," *Journal of Law, Medicine & Ethics* 34 (2006): 277–87.
30. Omin and Winant, *Racial Formation*; I. H. Lopez, *White By Law* (New York: New York University Press, 1996); M. A. Omi, "The Changing Meaning of Race," in *America Becoming*.
31. I. F. H. Lopez, "Race on the 2010 Census: Hispanics and the Shrinking White Majority," *Daedalus* 134 (2005): 42–52; K. Prewitt, "Racial Classification in America: Where Do We Go from Here?" *Daedalus* 134 (2005): 5–17.
32. A. Smedley, *Race in North America: Origin and Evolution of a Worldview* (Boulder, CO: Westview Press, 1999).
33. C. S. Fisher, M. Hout, M. S. Jankowski, and S. R. Lucas, eds., *Inequality By Design: Cracking the Bell Curve Myth* (Princeton, NJ: Princeton University Press, 1996).
34. Cooper, "Race and IQ: Molecular Genetics"; H. Gardiner, *Frames of Mind: The Theory of Multiple Intelligences* (New York: Basic Books, 1983); D. Goleman, *Working with Emotional Intelligence* (New York: Bantam Books, 1998); R. J. Sternberg, *The Triarchic Mind: A New Theory of Human Intelligence* (New York: Viking, 1986).
35. Anonymous, "Intelligence and Genetic Determinism"; Gould, *Mismeasure*.
36. Gardiner, *Frames of Mind*; Goleman, *Working with Emotional Intelligence*; Sternberg, *The Triarchic Mind*.
37. Gould, *Mismeasure*.
38. Ibid., 24.
39. R. J. Hernstein and C. Murray, *The Bell Curve: The Reshaping of American Life by Difference in Intelligence* (New York: Free Press, 1994); R. Lynn, *Dysgenics in Modern Populations* (Westport, CT: Praeger, 1996); R. Lynn and T. Vanhanen, *IQ and the Wealth of Nations* (Westport, CT: Praeger, 2002).
40. J. L. Kincheleo, S. R. Seteinberg, and A. D. Gresson, eds., *Measured Lies* (New York: St. Martin's Press, 1996); Fisher et al., *Inequality By Design*; M. K. Brown, M. Carnoy,

E. Currie et al., *Whitewashing Race: The Myth of a Colorblind Society* (Berkeley: University of California Press, 2003); R. Nisbett, *Intelligence and How to Get It: Why Schools and Cultures Count* (New York: Norton, 2009).

41. Ibid.
42. E. B. Isaacs, D. G. Gadian, S. Sabatini et al., "The Effect of Early Human Diet on Caudate Volumes and IQ," *Pediatric Research* 63 (2008): 308–14; T. I. Lidsky and J. S. Schneider, "Adverse Effects of Childhood Lead Poisoning: The Clinical Neuropsychological Perspective," *Environmental Research* 100 (2006): 284–93; Fisher et al., *Inequality By Design*; S. J. Ceci, "How Much Does Schooling Influence General Intelligence and its Cognitive Components? A Reassessment of the Evidence," *Developmental Psychology* 27 (1991): 703–22.
43. G. W. Evans and M. A. Schamberg, "Childhood Poverty, Chronic Stress, and Adult Working Memory," *Proceedings of the National Academy of Sciences* 106, no. 16 (2009): 6545–49.
44. Orfield and Lee, *Why Segregation Matters*; Fisher et al., *Inequality By Design*; S. R. Sirin, "Socioeconomic Status and Academic Achievement: A Meta-analytic Review of Research," *Revew of Educational Research* 75 (2005): 417–53.
45. J. S. Zax and D. I. Rees, "IQ, Academic Performance, Environment, and Earnings," *The Review of Economics and Statistics* 84 (2002): 600–16.
46. Fisher et al., *Inequality By Design*.
47. Lidsky and Schneider, "Adverse Effects."
48. Nisbett, *Intelligence and How to Get It*; J. R. Flynn, "The Mean IQ of Americans: Massive Gains 1932 to 1978," *Psychological Bulletin* 14 (1981): 623–28.
49. W. T. Dickens and J. R. Flynn, "Black Americans Reduce the Racial IQ Gap," *Psychological Science* 17 (2006) :913–20.
50. I. J. Deary and P. Smith, "Intelligence Research and Assessment in the United Kingdom," in *International Handbook of Intelligence*, ed. R. J. Sternberg, (Cambridge, Cambridge University Press, 2004).
51. Nisbett, *Intelligence and How to Get It*.
52. P. D. Evans, S. L. Gilbert, N. Mekel-Bobrov, E. J. Vallender et al., "Microcephalin: A Gene Regulating Brain Size Continues to Evolve Adaptively in Humans," *Science* 309 (2005): 1717–20; N. Mekel-Bobrov, S. L. Gilert, P. D. Evans, E. J. Vallender et al., "Ongoing Adaptive Evolution of ASPM, a Brain Size Determinant in Homo Sapiens," *Science* 309 (2005): 1720–22.
53. P. D. Evans, "Microcephalin."
54. Mekel-Bobrov et al., "Ongoing Adaptive Evolution of ASPM."
55. Gould, *Mismeasure*; Marks, *Human Biodiversity*.
56. M. Currat, L. Excoffier, W. Maddison et al., "Comment on 'Ongoing Adaptive Evolution of ASPM, a Brain Size Determinant in Homo Sapiens' and 'Microcephalin, a Gene Regulating Brain Size Continues to Evolve Adaptively in Humans,'" *Science* 313 (2006): 172(a); F. Yu, S. R. Hill, S. F. Schaffner, et al., "Comment on 'Ongoing Adaptive Evolution of ASPM, a Brain Size Determinant in Homo Sapiens,'" *Science* 316 (2007): 370(b).

57. P. Rushton, P. A. Vernon, and T. A. Bons, "No Evidence that Polymorphisms of Brain Regulator Genes Microcephalin and ASPM Are Associated with General Mental Ability, Head Circumference or Altruism," *Biology Letters* 3 (2007): 157–60.
58. Mekel-Bobrov et al., "Ongoing Adaptive Evolution of ASPM"; Evans et al., "Microcephalin"; T. C. Bates et al., "Recently-Derived Variants of Brain-Size Genes ASPM, MCHP1, CDK5RAP and BRCA1 Not Associated with General Cognition, Reading or Language," *Intelligence* 36 (2008): 689–693.
59. Bates et al., "Recently-Derived Variants," 689.

10

INTELLIGENCE, RACE, AND GENETICS

Robert J. Sternberg, Elena L. Grigorenko,
Kenneth K. Kidd, and Steven E. Stemler

A number of scholars claim to have studied relationships between intelligence, race, and genetics.[1] The thesis of this chapter is that many of these studies are not grounded in scientifically derived constructs but rather, in large part, in folk beliefs about them. There is a big difference between studying relationships between constructs and studying relationships between folk beliefs regarding those constructs. The bigger problem, however, is when one studies the latter but believes one is studying the former.

In the first part of this chapter, we review the constructs of intelligence and of race. In the second part, we discuss conceptual and methodological problems associated with studies that have attempted to examine the relationship between race, genes, and intelligence.

INTELLIGENCE

To study the interrelationships among intelligence, race, and genetics, we need to know what intelligence is. We do not know. Hence, any conclusions about its relationships to other constructs will be, at best, tentative.

Explicit Theories of Intelligence

One way to figure out what intelligence is has been to ask experts. Two major symposia have done so.[2] Each of the roughly two dozen definitions in each symposium was different. There were some common threads, such as the importance of adaptation to the environment and of the ability to learn. But these constructs themselves are not well specified. And very few tests of intelligence directly measure either one. Tests do not offer adaptive tasks one is likely to face in everyday life. Nor do any tests directly measure ability to learn, except dynamic tests that require learning at the time of test.[3] Rather, traditional tests much more measure past learning, which can have resulted from differences in many things, including motivation and available opportunities to learn.

Some theories of intelligence extend this definition by suggesting that there is a general factor of intelligence, often labeled *g*, which underlies all adaptive behavior.[4] In many theories, including the theories most widely accepted today,[5] other mental abilities are hierarchically nested under this general factor at successively greater levels of specificity. For example, Carroll has suggested that three levels can nicely capture the hierarchy of abilities, whereas Cattell and Vernon suggested that two levels were especially important.[6] In the case of Cattell, nested under general ability are fluid abilities of the kind needed to solve abstract reasoning problems such as figural matrices or series completions and crystallized abilities of the kind needed to solve problems of vocabulary and general information. In the case of Vernon, the two levels corresponded to verbal-educational and practical-mechanical abilities. What we know about group differences is largely about so-called *g* and major group factors, such as verbal and spatial skills. More modern theories extend intelligence much further, such as to creative and practical as well as analytical abilities[7] or to eight distinct multiple intelligences.[8]

Implicit Theories of Intelligence

Lay conceptions of intelligence are quite a bit broader than the conceptions of psychologists[9] who believe in general ability, or *g*.[10] For example, in a study of people's conceptions of intelligence Sternberg and his colleagues found that laypersons had a three-factor view of intelligence as comprising verbal, practical problem solving, and social-competence abilities.[11] Only the first of these abilities is measured by conventional tests. Experts in different occupations in the United States have somewhat different conceptions of intelligence, with their views of the relevant attributes tending to match the requirements of their occupations.[12] And conceptions of intelligence around the world vary even more than they do in the United States.[13]

The way intelligence is usually defined in studies of the alleged relationships between intelligence, race, and genetics is in terms of Boring's operational definition of intelligence as whatever it is that IQ tests measure.[14] This definition is unsatisfactory for at least three different reasons. First, the definition is circular, defining the construct in terms of the operation and the operation in terms of the construct. Second, so-called IQ tests do not all measure the same thing.[15] Third, as we have seen, theorists of intelligence do not themselves agree as to what intelligence is.

For convenience, we can follow Boring and operationally define intelligence in terms of IQ as measured by conventional tests. But it is not clear that tests of IQ measure the same construct among all the people to whom the tests are applied.[16] The more culturally distinct the people, the greater are the differences in what the items measure.[17] In part this is because IQ-test items are, largely, measures of achievement at various levels of competency.[18] Items requiring knowledge of the fundamentals of vocabulary, information, comprehension, and arithmetic problem solving—so-called measures of crystallized abilities[19]—are clearly measures of achievement. Items requiring fluid abilities[20] involving abstract reasoning, once thought to be culture-fair,[21] have proven even more susceptible to the effects of cultural and other environments than tests of crystallized abilities,[22] suggesting they are in no way "culture-fair." Western-style schooling even more extensively inculcates these ways of abstract or fluid thinking than it does those measured by tests of crystallized abilities.

In sum, it is probably more accurate to say that these existing studies refer to the relation between "IQ" or "psychometric *g*" and what is labeled as "race," rather than to "intelligence" and these other constructs. Does the language we use matter? Yes. We need to acknowledge that we are using convenient, partial operationalizations of the construct of intelligence, and nothing more. As professionals, some of us may understand that there is a large gap between the conceptualization and operationalization of intelligence. Others of us may act as though IQ tests somehow provide the kind of measurement of intelligence that a tape measure provides of height. When we are dealing with the lay audiences who learn about our work, it is especially important that we acknowledge that we have nothing even vaguely close to a "tape measure" of intelligence.

RACE

Just as there are different ideas about how to define and measure intelligence, there are several different ideas about how to define and measure race. Most scientists who study such matters believe that the original modern humans, of whom all living humans are descendants, lived in Africa.[23] They first appeared roughly two hundred thousand years ago. For whatever reasons—to find food, to satisfy wanderlust, to find better protection against predators, to find more land—small numbers of unrepresentative people started to migrate out of Africa about one hundred thousand years ago.[24]

The "Out-of-Africa" hypothesis places the first immigrants from Africa in southwest Asia. Over the course of tens of thousands of years, that initial non-African population expanded until there are now at least some people to be found on all continents and in most regions of those continents, except for Antarctica, which, in general, is too cold to be hospitable, at least for modern humans. As people migrated, they adapted to better fit their environments. Much of that adaptation was cultural—different clothing, different foods, for example—but some of the adaptation was genetic (e.g., a genetic response to the increased prevalence of malaria that occurred as a result of people's creation of agricultural fields and their irrigation). However, it is difficult to prove that traits seen to differ are truly the result of different selective pressures, that is, of genetic adaptations. A major reason for the difficulty is that at the genetic level we see quantitative differences in the frequencies of genetic variants, not qualitative genetic differences, among populations. When multiple forms of a DNA sequence, either a coding sequence or a noncoding sequence, are present, the sequence is referred to as polymorphic and the forms as alleles at the polymorphism. Among populations of various kinds, allele frequency differences at polymorphisms are the rule because of the chance effects known as "random genetic drift." In other words, as a result of both natural and social events, only some genotypes are transmitted through generations, whereas the others are lost; the lack of predictability in who will have children and who will not introduces powerful random noise into allele frequencies between generations. Thus, observing different allele frequencies does not in and of itself imply that local selection has operated. Even in the extreme cases of an allele absent in one part of the world and the only allele in another, we usually see a gradual difference (a cline) in allele frequencies in the populations along the geographic region between. One example is an allele at *EDAR*[25] that results in thicker hair.[26] The bottom line is that all the recent large-scale studies of human populations have concluded that genetic variation is clinal (i.e., gradual) around the world, with general loss of genetic variation in populations correlated with distance from Africa, much of this pattern a reflection of the way humans expanded throughout Eurasia, the Pacific, and the Americas in the last sixty to one hundred thousand years.[27]

MECHANISMS OF GENETIC INFLUENCE

Four mechanisms have influenced the genetic evolution of populations.[28] Consider each in turn. The first is *mutation*, by which DNA changes in random ways. Mutation results in the rise of both functional (i.e., coding) and nonfunctional (i.e., noncoding) polymorphisms as well as other structural variants.

The second is *random genetic drift*, by which alleles in finite populations may change in frequency over time as a result of the accumulation of random sampling error in the passing on of alleles from generation to generation. When a number of individuals migrates and starts a new population, the sampling error (random genetic drift) is inversely proportional to the number of founding individuals, and allele frequencies may be very different in the new population from those in the parent population. As the new population grows over a few generations, the magnitude of the sampling error per generation decreases and the new population will continue to have different frequencies from the parent population. The extreme form of random genetic drift is referred to as a "founder effect" because the population expanded from very few founders with a relatively restricted gene pool. For example, available evidence suggests that a small group of individuals left Africa, thereby changing the allele frequencies from those in the African populations left behind. On a smaller scale, the expansion of that population across Eurasia can be modeled as a series of smaller founder events resulting in gradual changes in frequencies along the paths of expansion.

The third mechanism is *gene flow* or *genetic exchange*, by which interbreeding among certain groups of individuals potentially results in those populations becoming increasingly similar to each other. Two populations that start off quite different genetically, if they mate, can produce offspring that represent the genes in both of the original populations. At a more local level, exchange between adjacent populations will, over time, smooth the geographic pattern into a smoother clinal gradient in frequencies.

The fourth mechanism is *natural selection*, by which organisms with gene patterns that are adaptive to a given environment become more prevalent over time. For example, organisms that can adapt to changing climatic patterns are at an advantage over those that adapt only with great difficulty.

MIGRATION AND ADAPTATION

Although all of these mechanisms are of importance, here we will illustrate only that of natural selection. Consider the following example. During the Industrial Revolution in late-nineteenth-century England, a particular dark-colored moth became more prevalent than a related light-color moth. Why? It is believed industrial pollution had blackened the forests and improved the darker moth's camouflage against predators such as birds. The light-colored moth was too visible to survive. More recently, however, with restrictions on air pollution, the light moth has made a comeback.[29] The point, of course, is that natural selection is a constantly shifting process. It is influenced not only by an organism's biology, but also by the interaction of that biology with environmental conditions.[30]

Is it better from the standpoint of adaptation to the physical environment to be a black moth or a light-colored moth? It depends on the interaction between the organism's attributes, including color, and the particular environment. Is it better from the same adaptive standpoint to be a black person or a light-colored person? The answer is the same, of course. In zones with more intense exposure to sunlight, darker skin puts individuals at an adaptive advantage. The melanin that acts as pigmentation to produce darker skin better protects individuals against the damage that large amounts of ultraviolet radiation can cause to the skin. Left unchecked, this radiation increases susceptibility to skin cancer, especially melanoma, a form of skin cancer that easily can become fatal. In zones with weaker exposure to sunlight, lighter skin is an advantage.

One explanation of lighter coloration pertains to vitamin absorption. People rely on sunlight to produce active vitamin D3 in the capillaries. The active form does not occur in great quantities in the food most people eat. Indeed, today milk is often supplemented with vitamin D3 to prevent deficiencies. Lighter skin allows greater bodily production of vitamin D3. Deficiencies in vitamin D3 can cause rickets in children and osteoporosis in adults.[31]

A second explanation is of a different kind. There is as yet no conclusive evidence for positive selection for light coloration. Instead, evidence to date may favor as much or more the simple relaxation in northern climes of the strong selection for dark pigment found in equatorial regions as an explanation for light coloration in zones distant from the equator.[32] Individual moths or other animals do not radically change in color in the course of their lifetimes. Rather, over time, those descendants that are better adapted are more likely to survive and reproduce, so that distributions of traits change. Human populations adapt over many generations. But not all organisms do. Some adapt very rapidly. Generations of bacteria, for example, adapt rapidly because of their extremely rapid rates of reproduction. It is for this reason that the same medication, Amoxicillin, which was effective in treating ear infections in the children of twenty years ago, is so much less effective in treating ear infections in the children of today. Bacteria have adapted, in the same way that malaria parasites have adapted to many quinine-based treatments and in the same way that the HIV virus is adapting to medications being used to treat it. All biological populations adapt, whether bacterial, human, or anything else.

There is another key fact in this story. Aside from the explanations of skin color, there are not a lot of scientifically supportable selective explanations for the differences we see in people from different parts of the world. It is probable that much of the variation that we see among groups of humans indirectly resulted from the pattern of expansion and migrations accompanied by random genetic drift. Over the years, frequencies of specific alleles at various single-nucleotide polymorphic sites (i.e., Single Nucleotide Polymorphisms, SNPs) changed only slightly in terms of nucleotide composition, but enough to make differences, many of which we still do not fully understand. The changes are numerous. Less than 1 percent of the three billion nucleotide positions in the human genome varies globally, as SNPs and other types of variation; but that percent creates a large number of potential differences between any two people. Some of the individual polymorphisms have different frequencies around the world; others have similar frequencies everywhere. In addition, structural variation, so-called Copy-Number Variation (CNV), can also be found; this type of variation has been discovered recently and has attracted much attention.[33] At this point, little is known about geographic differences in CNV but certain regions of the genome do seem prone to generate these duplications of genes or deletions of genes. The human genetic material, the genome, shows considerable variation among individuals when examined at the DNA level and the variants have different frequencies in different parts of the world. How that DNA variation affects variation in individual common traits is as yet poorly understood. Yet, we see larger proportions of blond hair and blue eyes in people born in European countries than those born in Asian ones. We will see shorter people, on average, among those born in Asia than among those born in Europe. We see wider noses in West Africa, on average, than in East Africa. Nevertheless, even within groups, there is tremendous variation.

RACE AS A SOCIAL CONSTRUCTION

Where does race fit into the genetic pattern? Actually, it fits nowhere. *Race is a socially constructed concept, not a biological one.* It derives from people's desire to classify. People seem to be natural classifiers. Perhaps this tendency reflects, in part, what Gardner has referred to as "naturalistic intelligence."[34] Or perhaps it merely reflects a need to discern order in or even to impose it on nature. Any set of observations can be categorized in multiple ways. People impose categorization and classification schemes that make sense to them and, in some cases, that favor their particular goals.

If one looks at geographic patterns, one will find many attributes that correlate with geography—nearby populations tend to be similar and distant populations dissimilar. This pattern is similar to common ideas of socially defined races but is more complex.[35] People in different places came to demonstrate different characteristics by adaptations to different environments, such as heterozygosity for sickle-cell hemoglobin as a partial protection against malaria, as well as by accumulation of random genetic drift. But as is so often the case, the same trait that may be adaptive in one circumstance may be maladaptive in another. For example, there is no advantage to sickle-cell hemoglobin in the absence of malaria and the anemia that results in homozygotic individuals poses a serious disadvantage.

Other adaptations are equally fickle. Today, our population is paying the price of tens of thousands of years in which people became genetically programmed to enjoy fats, sugars, and salt and to eat as much of them as they could when they have the opportunity. In the contemporary environment, the result is large numbers of overweight and obese individuals. Some people have more of a genetic predisposition to gain weight than others. Social stratification—classifying people into categories of higher and lower status in a society—has already ensued on the basis of weight.[36] Whether, ultimately, people with a genetic predisposition toward fatness will be classified as being of a separate race remains to be seen. The point is that an adaptation that is positive at one time or place may be indifferent at another and negative at still another.

One could pick any of a number of traits correlated with geographic patterns and find correlations with other related traits. It would be foolhardy, however, to view any one of these traits as causative of the others. That is what people have done who have viewed differences in so-called races as somehow causative of differences in IQ. It also would be foolhardy to group fairly arbitrary sets of traits and constructs that one then reifies as being natural, somehow God-given categories. One will find a distribution of traits in any of these groups, with only slightly more differentiation when comparing individuals from different groups rather than individuals within any one group.[37] Why would people do this, then? One reason is to justify existing social stratifications or to create new ones.

We could of course refer to moths as being of different "races" (black and white), in the same way we sometimes refer to humans as being of different "races." We do not typically use the term for moths, presumably because we are less interested in creating social stratifications for moths than for people, and race is one way to help create these stratifications. Of course, we recognize that our chapter may have the opposite effect from that intended: Some believers in biological race may realize that moths (and perhaps dogs, cats, and other animals that come in multiple colors) have been sorely neglected in the literature on racial differences, and that there is still time to remedy this situation. To the extent we define race as simply different sets of physical features, we could say, of course, that the moths are of different races. But the term, used in this way, becomes simply a word for saying the moths look different! And the surplus meaning associated with the word, at least as it is used in human descriptions, vanishes.

Over time, peoples who migrated changed both by chance and by adaptation to their environments in various ways. What is "good" depends on the adaptations that need to be made, and these adaptations change from time to time and place to place. For example, our ancestors in Africa were almost certainly dark-skinned because dark skin provided better protection against the particular challenges of the environment, most notably, ultraviolet and other harmful forms of radiation. Other traits, such as straight or curly hair, have no evident adaptive value and population differences probably reflect chance differences. Curiously, then, socially constructed judgments as to how socially to classify people are made on bases that have no relation to the original reasons that people came to look one way or another.

There is nothing special about skin color that serves as a basis for differentiating humans into so-called races. Any two groups of people that differ in one way are likely to differ in a cluster of ways. For example, as noted by Marks, geneticists have found that 54 percent of people who have designated themselves as Hebrew priests, many of whom have the surname Cohen, have a certain pattern of two genes on the Y chromosome.[38] In contrast, only 33 percent of Jews who do not view themselves as priests have this pattern. What conclusion is to be drawn? Well, the correct conclusion is that different groups of people will differ in various respects. The authors of the study concluded that one could infer a genetic Jewish priestly line dating back to the Biblical Aaron.[39] Other bases for differentiation could be chosen as well, including the aforementioned one of girth. The point is that people will often draw conclusions that go well beyond the data, as when they take a correlation to imply causation or when they take genetic variation to have implications for a Jewish priestly line. There may be a causal link, but the evidence is insufficient to conclude as such.

As another example, Fish has pointed out that people who have lived over many generations in cold climates, such as Eskimos, have tended to develop rounded bodies to maintain heat and thus stay warmer.[40] Some populations in very hot climates, such as the Masai, have tended instead to develop lanky bodies. The hypothesis is that the high ratio of surface area to volume results in their radiating a lot of heat and thus staying cooler. While reasonable, both adaptation hypotheses lack rigorous scientific proof. Possibly, they could be just coincidences. Scientists do not know for sure.

In the American folk taxonomy of race, as Fish argued, lanky and rounded people can be, respectively, two kinds of blacks and whites.[41] But one could as easily decide that a more "basic" taxonomy of races would be in terms of lanky and rounded bodies, in which case there would be black and white members of the lanky and rounded races. One would find a number of genetic patterns that, on average, correspond to lankiness and roundedness, in the same way one would find genetic correlate patterns corresponding to darker or lighter skin, or Cohens versus non-Cohens, or basketball players versus wrestlers.

It has been argued that the challenges faced by those who migrated to Northern climates were greater than those faced by people in Southern climates, and that this difference might have led to higher intelligence of those who went northward.[42] However, anyone who has spent any significant time in Africa might well dispute this claim. One of the greatest challenges of tropical climates is fighting tropical diseases to survive, and the challenges of fighting diseases are greater in the tropics than they are further north. Indeed, children acquire from an early age specialized knowledge, not acquired further north, regarding natural herbal medicines that can be used to combat tropical illnesses.[43] To the extent that warmer climates encourage greater aggression,[44] learning how to compete successfully so as to survive in such environments also might promote intellectual development. We are not arguing that people in warmer climates did indeed develop higher intelligence, but rather, that one could create speculative arguments supporting greater intellectual growth in such climates, as has been done to support the notion that there was greater intellectual growth as a result of challenges up north. Indeed, post hoc evolutionary arguments made in the absence of fossils at times can have the character of ad hoc "just so" stories designed to support in retrospect whatever point the author wishes to make about present-day people.

Differences in socially constructed races stem largely from geographic dispersions that happened long past, starting about one hundred thousand years ago but continuing until about three thousand years ago in some areas. Today we see the physical correlates left by the dispersions. Much of that variation is continuous across distances but with different traits showing different rates and patterns of change. What "race" does is to reify these differences as deriving from some imagined natural grouping of people that does not, in fact, exist, except in our heads.

What we see in terms of skin color correlates very well with our developed folk taxonomies, but only weakly with genetic differentiations. For example, the amount of genetic variation in Africa is enormous and is much greater than that in the rest of the world.[45] In contrast, in terms of the amount of phenotypic variation, or differences in appearances, Africa is at least comparable to the rest of the world. The phenotypic differences are nevertheless notable. For example, in Africa, one can find very tall Masai, and very short Pygmies who probably gained an adaptive advantage by virtue of their shortness for locomotion through forest vegetation.[46] Yet, some may lump together all these Africans as the same, though genetically they differ more from each other, in many cases, than they do from those who perceive themselves to be of higher social, or even biological, value.

Humans have devised various metaphors for understanding why some people are more successful, according to whatever standards society invents, than others. Usually, the comparisons are drawn by those who consider themselves successful for the benefit of others who consider themselves successful, or on the road to success.[47] A curiosity of history is that people come to believe in the reality of their own metaphors. For example, some have believed, and some still believe, in "royal blood." Educated people probably realize that the expression is metaphorical; others probably believe that the blood of royals differs in some key respect from the blood of others.

For readers of this chapter, a biological concept of "royal blood" probably seems silly. But at the same time, we know that there are distinguishing blood groups. For example, most of us are familiar with the ABO and Rh blood-typing systems. According to Lewontin, there are roughly thirty-five blood group systems, with fifteen serving at least somewhat effectively to distinguish different human populations.[48] Royal blood, at least within families, may well show distinguishing blood groups, just as nonroyal families would. So in this trivial sense, royal blood can be said to exist, but differently in different royal families. In this same trivial sense, there can be differences in distributions of blood groups across religious groups, people with different body shapes, or people with different skin colors.

How mixtures are labeled is a function of social status. In the United States, blacks generally have lower social status than whites, so supposed admixtures of blood determine degrees of "blackness." Having any blackness makes one socially black in some degree. So one can be light black, or medium-skinned, or dark black, but one is still socially black. Even if one of mixed parentage inherited none of the physical features of blackness, one would still be classified socially as black, although one might pass for white.[49] Where blacks are of higher social status, degrees of whiteness may all be seen as departures from true blackness.

The concept of race serves a social, not a biological, purpose. Different types of parentage have, at various times and places, given rise to racial labeling, as, for example, in the "Aryan race," the "German race," the "Jewish race," and so forth. In Apartheid South Africa, the races were Bantu (Black African), colored (including people of perceived mixed descent), Indian/Asian, and white. In contemporary North American society, we mix together the black and colored "races," somehow believing, as noted above, that if someone has any degree of nonwhiteness, it puts that individual into the black category. Hitler designated as a member of the Jewish race anyone who had supposed Jewish blood, which could date back to one's great-grandparents.

In Brazil, the supposed races are different again.[50] A *loura* has straight blond hair, blue or green eyes, light skin color, and a narrow nose and thin lips. A *branca* has light skin color, eyes and hair of any color, a nose that is not broad, and nonthick lips. In Brazil, Fish points out, a *branca* is white.

In the United States, a *branca* individual from Brazil would more likely be classified as "Hispanic." Then there is a *morena*, who has brown or black hair that is wavy or curly but not tight curly, tan skin, a nose that is not narrow, and lips that are not thin. *Morenas* in the United States are classified as black or Hispanic. There are a number of other Brazilian terms used to describe socially constructed racial categories, such as *mulata* and *preta*, and to the Brazilians, these terms are every bit as real as the black, white, and Asian categories are in the United States. They *are* real. But as in the United States, they are folk, not biological, taxonomies, which are used to socially stratify people, often in the name of science. At best, the effects are innocuous. At worst, they become the bases of genocides.

People generally use skin color to distinguish races, but not always. During the genocide in Rwanda, the Hutus used other physical attributes, such as height, to distinguish Tutsis. Because there had been so much intermarriage between Hutus and Tutsis, the distinctions were generally weak, and many people were killed simply because they seemed closer to the imagined Tutsi prototype than the Hutu one, regardless of their origins.

The history of the concept of race is inextricably intertwined with attempts by the winners to explain or justify why they perceive themselves to be winners. Consider, for example, the term "Caucasian." It is an odd term, in some ways, because although it is used to refer to "whites," in Russia, people from the Caucuses are considered dark relative to many other Russians. Especially because of political difficulties in Chechnya and surrounding areas, these dark Caucasians today are viewed with suspicion and distrust in much of Russia. So the term that is accepted as "scientifically" identifying white people in the United States, often in preference to the term "white" to give more of a feeling of scientific classification, is used in a way that is largely opposite in contemporary Russia. Where did the term come from then? It was invented in 1795 by Johann Friedrich Blumenbach,[51] who chose the name because he believed that the Georgians, from the Mount Caucasus region, were the most beautiful race of men (his words). The term stuck. So people in English-speaking countries with white skin have the honor of having a name they imagine to be the formal name for their race representing what one naturalist in 1795 believed was the most attractive "race" and what today largely is believed to be rather dark, not white, skin according to Russian standards. Thus, the term is scientifically unsupportable and part of an old racist typology. The term is just as racist as Negroid and Mongoloid, terms the politically sensitive American will not use.

ORIGINS OF THE CONCEPT OF RACE

Whence emerged the concept of "race"? The concept of race as a classification scheme representing allegedly natural "types" distinguishable on the basis of clear visual attributes such as skin or eye color, hair texture, and certain facial and bodily features was initially introduced in the seventeenth century.[52] However, it took these ideas almost a century to attract the attention of scientific "authorities." According to Gould, Linnaeus (in 1758) first proposed four races: *Americanus*, *Europaeus*, *Asiaticus*, and *Afer* (or African).[53] He also alluded to two other categories that did not prove as useful for social purposes as the other categories: wild boys (feral children) discovered in the forests, and monsters, hairy men with long tails, who emerged from tales of travelers. Blumenbach (in 1775), building on the work of Linnaeus, first proposed a grouping of "races," namely, Caucasians, Mongolians, Ethiopians, and Malays. This early history was no more scientific than the later history was to be. That is, race started out as a not so subtle way of socially classifying and, ultimately, stratifying people hierarchically—as better or worse. For example, Linnaeus viewed the white as sanguine and muscular, and the black as phlegmatic and relaxed.

Historically, the formation of the concepts of race and ethnicity was influenced by two main perspectives.[54] One perspective was formed in the context of the eugenics movement and was used to refer to presumed biological differences between socially defined populations.[55] The other perspective was formed in the context of physical anthropology and the social sciences and rejected the idea of the biological significance of racial classifications. It argued that race and ethnicity are primarily cultural and historical products of human history.[56] Today, whereas some still defend the basis for the "gene-based evolutionary theory" of race,[57] the majority of cultural anthropologists are in agreement that race is a socially constructed, not an evolutionary determined or biologically supported, concept.[58] Of course, science does not find truth by majority rule. *The problem with the concept of biological race is not that it is supported by only a minority of anthropologists, but that it has no scientific basis.* Moreover, attempts to link intelligence, race, and genetics have also lacked adequate scientific foundation.

INTELLIGENCE, RACE, AND GENETICS

Despite the inadequacies that have been pointed out with regard to the definitions of both race and intelligence, several studies have attempted to examine the relationship between race and intelligence using proxies that are intended to intimate a biological basis for each construct. One set of studies attempts to give biological credence to the concept of race by equating it with skin color. Another set of studies attempts to justify the reification of race using the argument that race is an important factor in customizing medical diagnoses and treatments. A third set of studies uses correlational twin study data to make causal assertions about the relationship between intelligence, genes, and race. A fourth set of studies relies on misguided interpretations of heritability studies for making cross-racial comparisons. Each of these arguments suffers from a number of serious flaws.

Skin Color Is Not Tantamount to Race

Many studies that purport to investigate race as a biologically based construct use self-reported skin color as a proxy for some sort of presumed innate biological marker of race.[59] There are several problems with this approach. First, the operational definition of race based on skin color varies substantially over time and space. What one group sees as one race based on a certain color (e.g., "black" in the United States), another group may see as another race based on the same color (e.g., "colored" in South Africa).

Second, even if self-reported skin color could be reliably measured over time and space, there is no genetic evidence to support the idea that individuals with a shared skin color share other types of genes in common more frequently than individuals of different skin colors. A simple thought experiment will illustrate this point. Suppose we put one hundred randomly selected people in a room together behind a curtain. We then have a competition in which the object of our game is to choose the individuals with the closest genetic match. Contestant A and Contestant B each only get one variable on which to evaluate those one hundred individuals. Contestant A chooses skin color as his variable. Thus, the forearms of one hundred individuals are stuck through the curtain and individuals are categorized into groups according to the degree to which their skin colors are similar. Contestant B chooses to listen to each of the one hundred individuals say a sentence in English. The individuals who sound the most alike (indicating that they are from the same part of the world, most likely, even if they differ drastically in their skin color) are then categorized into groups. Chances are that Contestant B will win the game much more often than Contestant A if we run an analysis of all of the alleles these two individuals in both groups share. The individuals who sound more alike are more likely to hail from the same parts of the globe and will therefore be more genetically similar to each other. But certainly similarity in English speech patterns does not form the basis for a racial classification.

The point is that although skin color is genetically determined, it does not imply that people with the same skin color share many other genes in common. In fact, there is a tremendous amount of variability in the extent to which two individuals with the same skin color share their remaining genes.[60] By contrast, the genetic evidence does suggest that individuals from a similar part of the world tend to share more genes in common than people from parts that are remote from one another. The more geographically distant individuals are from each other, the fewer genes they seem to have in common, on average.

Third, the data presented by Templer and Arikawa[61] and by Lynn[62] showing correlations between "national skin color" and "national average IQ" suffer from many conceptual and methodological problems that have been addressed in detail by others in the literature.[63] One of the more blatant problems with these data is that the samples used are not random selections from the population, but rather samples of convenience. Perhaps the most basic lesson of survey methodology is that sample size is no substitute for sample representativeness. Although a truly representative sample of approximately three thousand individuals can reasonably accurately represent three hundred million individuals, even a sample size of over two million individuals, when not representatively sampled, can lead to gross errors in statistical inference due to the infiltration of uncontrolled third variables. This finding was perhaps most notoriously illustrated by the 1936 US election in which *Literary Digest*, the top pollster in the United States at the turn of the century, conducted a poll of over two million individuals and predicted that Alf Landon would defeat Franklin Roosevelt in a landslide. The result, of course, was just the opposite. FDR defeated Landon in a landslide. The problem was that the pollsters did not recognize the confounding variables that crept into the study when they drew their sample from lists of car owners and telephone owners in 1936 and they relied on voluntary response (just over 20 percent of individuals who were mailed surveys responded). Thus, any data from nonrepresentative samples that are used to make inferences to a larger population are suspect.

A second major problem with these data is that they concocted a measure of "national skin color" using the opinions of three graduate students. Not only does the concept of an "average" national skin color ignore important variability within each nation with regard to skin tone differences, but the fact that three independent students agreed on these classifications suggests only that they share the same implicit theories, prejudices, erroneous preconceptions, etc. Thus, what is being measured is not "national skin color" but rather social stereotypes about skin color.

In addition to those critiques, however, there is also a statistical-inference problem known as the ecological fallacy that plagues such data. The ecological fallacy occurs when relationships observed in groups are assumed to hold for individuals.[64] Cross-level inferences are not supportable without running multilevel statistical models.[65] Kreft and de Leeuw illustrate the problem of cross-level inferences when they note that:[66]

> Kreft et al. . . . ran a study in which data were collected on workers in 12 different industries. Individual-level variables are education level as the explanatory variable, and income as the response variable. The type of industry, as well as the distinction between public and private industries, are the second-level variables. An analysis with these data, executed at the level of individual workers, shows a positive relationship between education level and income: the higher the educational level, the higher the personal income. An analysis executed at the higher level, the industry level, with 12 industries as observations, shows a surprisingly opposite result. A negative relationship shows up between education and income.

> The higher the average educational level of an industry, the lower is the average income of workers in that industry. Universities and colleges are a good example.

In a similar vein, investigations that occur at the group level (e.g., "group intelligence" and "national skin color"), which are then used to make individual-level inferences, cannot be supported without multilevel statistical models.

Racial Profiling in Medicine

One of the other major arguments found in the literature suggests that race must have a biological basis because the field of medicine is currently actively pursuing racial profiling for medical treatments. For example, some research exists to show that different ethnic groups in the United States exhibit substantial average differences in areas such as disease incidence, prevalence, severity, and response to treatment.[67] Furthermore, there is ample evidence to suggest that the health disparities observed between different ethnic groups in the United States arise mostly through the environmental effects of discrimination, poverty, restricted access to health care, stress, and other socially mediated forces.[68]

It is worth noting that in February 2001 the editors of the medical journal *Archives of Pediatrics and Adolescent Medicine* asked authors no longer to use race as an explanatory variable and not to use obsolescent terms.[69] Some other high-impact peer-reviewed medical journals, such as the *New England Journal of Medicine* and the *American Journal of Public Health*, have made similar appeals.[70]

In the end, perhaps the fundamental problem with racial profiling in medicine is that it ignores the importance and potential confounding of environmental influences. As Risch et al. point out, "The true complication is due to the fact that racial and ethnic groups differ from each other on a variety of social, cultural, behavioral, and environmental factors as well as gene frequencies, leading to confounding between genetic and environmental risk factors in an ethnically heterogeneous study."[71] In other words, even if one could be confident about true genetic differences between groups and if one could create customized drugs for particular populations, one would need to take into account the effects that systematic cultural differences make with regard to areas such as diet and exercise.

Twin Studies

Perhaps the most widely used approach historically to studying the link between intelligence, race, and genes has been through the use of twin studies. After all, the argument goes, identical twins share 100 percent of their genes in common, siblings share 50 percent of their genes in common, and strangers share 0 percent of their genes in common. Thus, by studying each combination reared in the same environment or reared apart, we should be able to firmly disentangle the influence of genes from the influence of environment on variables such as intelligence. Indeed, several studies have demonstrated that over time, the correlations between the IQ scores of identical twins become stronger whereas the correlations of IQ scores among siblings shrink to nearly zero.[72] On the surface, these data appear to provide a powerful argument in favor of the influence of genes upon IQ. Unfortunately, it is not so simple to disentangle genetic influences from environmental influences this way.

For example, as Flynn has pointed out, for us to make a genetic attribution to the high correlation in IQ among identical twins, we must assume that their environments have no more in common than those of randomly selected individuals.[73] Yet, this assumption is likely to be untenable. Indeed, small genetic differences can interact with the environment to lead to what are called "multiplier effects." As an example, Flynn asks us to consider an analogy. Suppose a pair of identical twins is separated at birth. Both individuals may live in a social context that values a sport like basketball. And both twins may have a slight genetic advantage that makes them taller and quicker than average children. As a result of this slight advantage, both individuals will most likely be selected at an early age to play on the local basketball team. They may then receive more practice, coaching, and playing time, which then influence their basketball playing ability. The effects of these increased opportunities for further exposure to the game and development of skills lead both individuals to demonstrate strong basketball skills as they age. By contrast, consider a second set of identical twins separated at birth who are genetically shorter and chubbier than the average child. These children would likely not be selected for the basketball team and would not be exposed to the increased hours of practice, etc. As a result, these twins would also score similarly poorly on a test of basketball ability when they age. Thus, "genetic advantages that may have been quite modest at birth have a huge effect on eventual basketball skills by getting matched with better environments—and genes thereby get credit for the potency of powerful environmental factors."[74] Thus, one way to explain the increased similarity is via multiplier effects. A second explanation for the recent data by Davis et al. is that as people age, they gain more direct control over their environment and are therefore better able to select environments that are aligned with their genetic predispositions.[75] For example, more athletic children may turn into adults who like to spend most of their time outside working whereas children who may not be as athletic may enjoy spending more time reading. Thus, as individuals age, they will tend to choose environments that enhance their strengths, and these environments may differ from individual to individual.

Race and Heritability Studies

The explosion of genetic research within the last ten to fifteen years has brought the concept of race back to the surface, with some researchers arguing that new molecular data have given the concept of race new significance in the context of medicine and public health.[76] One might think that, because the concept of race originated as a social proxy for the description of biological differences, at least the biologists studying race would agree on its definition. However, the reality is different. When variation in genetic markers or allelic variants is considered, opinions range widely. One view is that socially defined racial differentiation is most pronounced and even discontinuous when it is evaluated on the basis of continental residence.[77] A second view is that there is continuity in genetic variation across socially defined races and that various races are not distinct, but rather a single lineage with a shared evolutionary fate.[78] On this view, there is no biological value in the concept of race.[79] However, considering these positions, it is important to understand that, even within these extreme views, researchers agree that, although human populations might differ dramatically in terms of proportions or frequencies of alternative forms of genes, that is, allelic variants, they do not differ in the kinds of genes they possess.[80] In fact, both extreme views may have some merit.[81]

A key argument of this article is that race is every bit as real as royal blood. It exists in some trivial sense as a correlate of various biological groupings stemming from migration and breeding patterns, and no more. However, just as royal families are usually interconnected and difficult to partition off fully, defining the boundaries between races is impossible. As *The American Heritage Dictionary of the English Language* notes on usage, "many cultural anthropologists now consider race to be more a social or mental construct than an objective biological fact."[82]

Although attempts have been made to establish genes for intelligence, no single gene has been conclusively identified.[83] To date, there have been six genome-wide scans for genes contributing to intelligence and cognition.[84] The results of these scans vary, but there are interesting partial overlaps. Specifically, the findings coincide in regions on chromosomes 2q (for four out of six studies), 6p (for five out of six studies), and 14q (for three out of six studies). These overlapping regions have been tentatively interpreted as indicative of the presence of genes that could explain some of the variance in IQ. Further, research has shown that specific genes such as *APOE*, *COMT*, and *BDNF* may play a role in intelligence; however, an in-depth understanding of the role of these genes remains elusive.[85] The IQ QTL project—a project aimed at identifying quantitative trait loci (QTL) contributing to genetic variation in intelligence[86]—has attempted to establish QTLs associated with intelligence, but to date, whatever positive findings have emerged have either failed to replicate,[87] or produced weak signals that have not yet been attempted to be replicated in independent samples.[88] Most recently, Deary et al. found that "there is still almost no replicated evidence concerning the individual genes, which have variants that contribute to intelligence differences."[89] Of course, the future may bring conclusive identifications: we just do not know yet.

As a result, virtually all attempts to study genes related to intelligence have been indirect, through studies of heritability. But heritability is itself a troubled concept. Are differences in intelligence between so-called races heritable? The question is difficult to answer in part because it is difficult even to say what can be concluded from the heritability statistic commonly used. Consider some facts about heritability.[90]

What Heritability Tells Us

Heritability (also referred to as h^2) is the ratio of genetic variation to total variation in an attribute *within* a population. Thus, the coefficient of heritability tells us nothing about sources of between-population variation. Moreover, the coefficient of heritability does *not* tell us the proportion of a trait that is genetic in absolute terms, but rather, the proportion of variation in a trait that is attributable to genetic variation within a specific population.

Trait variation in a population is referred to as phenotypic variation, whereas genetic variation in a population is referred to as genotypic variation. Thus, heritability is a ratio of genotypic variation to phenotypic variation. Heritability has a complementary concept, that of environmentality. Environmentality is a ratio of environmental variation to phenotypical variation. Note that both heritability and environmentality apply to populations, not to individuals. There is no way of estimating heritability for an individual, nor is the concept meaningful for individuals. Consider a trait that has a heritability statistic equaling 70 percent; it is nonsense to say that the development of the trait in an individual is 70 percent genetic.

Heritability is typically expressed on a 0 to 1 scale, with a value of 0 indicating no heritability whatsoever (i.e., no genetic variation in the trait) and a value of 1 indicating complete heritability (i.e., only genetic variation in the trait). Heritability and environmentality add to unity (assuming that the error variance related to measurement of the trait is blended into the environmental component). Heritability tells us the proportion of individual-difference variation in an attribute that appears to be attributable to genetic differences (variation) within a population. Thus, if IQ has a heritability of .50 within a certain population, then 50 percent of the variation in scores on the attribute within that population is due (in theory) to genetic influences. This statement is completely different from the statement that 50 percent of the attribute is inherited.

An important implication of these facts is that heritability is *not* tantamount to genetic influence. An attribute could be highly genetically influenced and have little or no heritability. The reason is that heritability depends on the existence of individual differences. If there are no individual differences, there is no heritability (because there is a 0 in the denominator of the ratio of genetic to total trait variation in a given population). For example, being born with two eyes is 100 percent under genetic control (except in the exceedingly rare case of severe dismorphologies, with which we will not deal here). Regardless of the environment into which one is born, a human being will have two eyes. But it is not meaningful to speak of the heritability of having two eyes, because there are no individual differences. Heritability is not 1: it is meaningless (because there is a 0 in the denominator of the ratio) and cannot be sensibly calculated.

Consider a second complementary example, occupational status. It has a statistically significant heritability coefficient associated with it,[91] but certainly it is not under direct genetic control. Clearly there is no gene or set of genes for occupational status. How could it be heritable, then? Heredity can affect certain factors that in turn lead people to occupations of higher or lower status. Thus, if things like intelligence, personality, and interpersonal attractiveness are under some degree of genetic control, then they may lead in turn to differences in occupational status. The effects of genes are at best indirect.[92] Other attributes, such as divorce, may ran in families, that is, show familiality, but again, they are not under direct genetic control; in fact, the familiality may be because they are culturally "inherited."

Heritability Can Vary Within a Given Population

Heritability is not a fixed value for a given attribute. Although we may read about "the heritability of IQ,"[93] there really is no single fixed value that represents any true, constant value for the heritability of IQ or anything else, as Herrnstein and Murray and most others in the field recognize.[94] Heritability depends on many factors, but the most important one is the range of environments. Because heritability represents a proportion of variation, its value will depend on the amount of variation. As Herrnstein pointed out, if there were no variation in environments, heritability would be perfect, because there would be no other source of variation.[95] If there is wide variation in environments, however, heritability is likely to decrease.

When one speaks of heritability, one needs to remember that genes always operate within environment contexts. All genetic effects occur within a reaction range, so that, inevitably, environment will be able to have differential effects on the same genetic structure. The reaction range is the range of phenotypes (observable effects of genes) that a given genotype (latent structure of genes) for any particular attribute can produce, given the interaction of environment with that genotype. For example, genotype sets a reaction range for the possible heights a person can attain, but childhood nutrition, diseases, and many other factors affect the adult height realized. Moreover, if different genotypes react differently to the environmental variation, heritability will show differences depending on the mean and variance in relevant environments.[96] Thus, the statistic is not a fixed value. There are no pure genetic effects on behavior, as would be shown dramatically if a child were raised in a small closet with no stimulation. Genes express themselves through covariation and interaction with the environment, as discussed further later.

Heritability and Modifiability

Because the value of the heritability statistic is relevant only to existing circumstances, it does not and cannot address a trait's modifiability. A trait could have zero, moderate, or even total heritability and, in any of these conditions, be not at all, partially, or fully modifiable. The heritability statistic deals with correlations, whereas modifiability deals with mean effects. Correlations, however, are independent of score levels. For example, adding a constant to a set of scores will not affect the correlation of that set with another set of scores. Consider height as an example of the limitation of the heritability statistic in addressing modifiability. Height is highly heritable, with a heritability of over .90. Yet height also is highly modifiable, as shown by the fact that average heights have risen dramatically throughout the past several generations.

As an even more extreme example, consider phenylketonuria (PKU). PKU is a genetically determined, recessive condition that arises due to a mutation (or, rather, a number of various rare mutations resulting in similar functional damages to the coded protein, see below) in a single gene, the *PAH* gene, on chromosome 12 (with a heritability of 1), and yet its effects are highly modifiable. Feeding an infant with PKU a diet free of phenylalanine prevents the mental retardation that otherwise would become manifest. Note also that a type of intellectual disability that once incorrectly was thought to be purely genetic is not. Rather, the intellectual disability associated with PKU is the result of the interaction with an environment (a "normal" diet) in which the infant ingests phenylalanine. Take away the phenylalanine and you reduce level of, or, in optimal cases, eliminate intellectual disability. Note that the genetic endowment does not change: the infant still has a mutant gene causing phenylketonuria. What changes is the manifestation of its associated symptoms in the environment. Similarly, with intelligence or any other trait, we cannot change (at least with our knowledge today) the genetic structure underlying manifestations of intelligence, but we can change those manifestations, or expressions of genes in the environment. Thus, knowing the heritability of a trait does not tell us anything about its modifiability.

Within-Population Effects Versus Between-Population Effects

One of the worst intellectual slips that have been made by investigators of heredity and environment (or rather, most often, by interpreters of findings on heredity and environment) is to generalize the effects of within-population studies between populations. For example, some investigators have made attributions about effects of racial or ethnic group differences on the basis of behavior-genetic studies, even while admitting that such conclusions are sometimes flawed.[97] All of the behavior-genetic designs in the studies noted above can ascertain effects of genetic variation only within populations. For example, they may tell us something about the extent to which individual differences in the measured intelligence of people in a particular group are associated with genetic factors. They say nothing about sources of between-population differences in levels of measured intelligence.

An illustration of the impossibility of making between-population claims from within-population data has been given by Lewontin.[98] Specifically, in a study using a set of protein markers (blood groups, serum proteins, and red blood cell enzymes) as indicators of genetic differences between populations, Lewontin estimated that roughly 85 percent of the genetic variance occurs between any two individuals within any socially identified racial groups, roughly 9 percent occurs among different populations within a socially identified race, and only the remaining 6 to 7 percent occurs between socially identified races. Other researchers arrived at the same conclusions using more powerful datasets obtained with more technologically advanced methodologies[99] or through simulation analyses.[100]

Different populations—racial, ethnic, religious, or whatever—may encounter quite different environments, on average. Whatever the heritability of intelligence or other attributes within a given setting, no conclusions can be drawn about heritabililty as a source of differences across settings. The fact that IQs have increased so much over the years suggests that environments differ widely over time.[101] They likely differ substantially as well for members of different groups at a given time.

Nisbett reviewed published studies investigating sources of differences in cognitive abilities between white and black individuals.[102] These studies, using designs unlike the behavior-genetic studies described above, have directly sought to investigate genetic and environmental effects on intelligence. For example, one design (Scarr and Weinberg) has been to look at black children adopted by white parents. Of seven published studies, six supported primarily environmental interpretations of group differences, and only one study did not; the results of this one study are equivocal.[103] What the Scarr and Weinberg work study did show is that IQs of adopted children are more similar to those of their biological mothers than to those of their adopted mothers. Less clear are the "racial" implications of their findings.

Moreover, there is much published evidence indicating that heritability estimates vary across populations. For example, estimates of the heritability of IQ in Russian twin studies conducted in the Soviet era tended to be higher than comparable estimates in the United States.[104] This observation made sense: environmental variation in Russia under the Soviet regime was constrained; consequently, heritability estimates were higher. Most of the IQ heritability studies up to today have been carried out in various countries of the developed world. Relatively little information exists regarding the heritability of IQ in the developing world, although some studies suggest that heritability may be substantial, at least outside the Western countries that most often have been studied.[105]

In sum, heritability estimates do not explain the genetic regulation of behavior and do not provide accurate estimates of the strength of the genetic regulation. Heritabilities are like snapshots of a dancer. Heritabilities will not tell us either what the dance is about or what is coming next in the dance. The true genetic nature of humans is far from being defined.

But what is absolutely clear is that genes do not act in a vacuum; they act in the environment, and their actions can be altered by the environment.

BIOLOGICAL AND GENETIC DATA AS RELATED TO THE CONCEPT OF RACE

One would hope that, because the concept of race was originally, if falsely, conceived as a concept to signify the degree of biological differences between groups of people, the strongest support for the concept of race would originate from biological and genetic data. Does it? Here we review some examples of the relevant research. First, it appears that the global distribution of genetic variation in humans is not easily sorted by so-called races. As reviewed recently, scientists have studied diverse populations for many polymorphisms.[106] These studies involve polymorphisms in the nuclear DNA, including variation in the non-recombining Y chromosome and autosomal (i.e., located on chromosomes other than Y and X) markers as well as polymorphisms in the mitochondrial DNA.[107] A clear consensus picture has emerged of the distribution of genetic variation around the world, at least in broad strokes. These data overwhelmingly support the following model for recent human evolution and diversification of populations.

Modern *Homo sapiens* evolved once in Africa about two hundred thousand years ago and then spread throughout the rest of the world and simultaneously diversified starting about fifty to one hundred thousand years ago. During that spreading out, modern humans supplanted now-archaic humanlike populations identifiable as having spread outside of Africa, such as Neanderthals.The evidence is that effectively only one population left Africa and settled in southwest Asia. That population was characterized by a large founder effect before it expanded into other regions. From that population, different pathways of expansion occurred: into Europe and separately across Asia. At some point in Asia, not yet clearly identifiable, additional expansions occurred, one expansion into northeast Asia and then into the Americas, plus a separate expansion into Melanesia and Australia. Associated with all of these expansions is accumulating random genetic drift at all polymorphic sites of the genome. Thus, allele frequencies generally show gradual changes as one moves around the world. Of course, recent migrations (over the last few thousand years) of established populations into already-occupied regions can result in some adjacent populations having very different allele frequencies, but that has been rare until historic times. Today in the United States, for example, we have populations from very different parts of that geographically continuous spectrum of allele frequencies. Those distinct allele frequencies do not mean that different "races" exist, only that different parts of a continuum have been sampled. An analogy is the distinction between the colors blue, yellow, and red as samples from a continuous spectrum of light. Those colors only have meaning because the spectral sensitivities of the photoreceptors in our eyes and the neurological circuits interpreting the signals interact with a label arbitrarily imposed on some narrow range of wavelengths from a continuous spectrum.

There is no question that populations, defined geographically, demonstrate dramatic variability in frequencies, not only for the several million normal polymorphisms not associated with causing genetic disorders but also for many disease-related genetic alleles (variants). The genetic alleles (variants) can be readily seen in ALFRED, the Allele Frequency Database.[108] The issue is not whether this variation is present or not; the issue is whether explaining this variation should occur at the levels of populations per se (e.g., Lapps, Chuvash, Nyanja, or Corsicans), continents (e.g., Europe or Africa), or alleged races. After our review of the literature, we think that variation that seems to be meaningful and transferable into helpful public health or educational policies is at the level of specific populations. Global socially constructed categories such as race do not appear to be useful proxies for genetic features.

Second, considering evidence for a biological basis for racial classification, it is important to appreciate comparatively the amount of genetic variation observed within and among specific racial categories. In this context, let us turn for an illustration to the research on genetic bases of complex diseases. From rapidly accumulating evidence, it seems that a number of geneticists have stated that most common complex diseases, such as diabetes, hypertension, cancer, and so forth, appear to be at least partially governed by genetic mechanisms, shared by most, if not all, populations around the world.[109] This statement has triggered a number of large-scale studies such as projects in Iceland and Estonia, where population-wide genome banks have been created in the hope of identifying specific alleles associated with common diseases within populations so major pathways of genetic disorders can be discovered and later generalized to other populations.[110] Although the effectiveness of this approach is yet to be determined, this approach has been encouraged by new evidence indicating that many uninterrupted or rarely interrupted chunks of DNA (referred as haplotypes) appear to be common across different populations socially classified as belonging to different races.[111]

To appreciate the significance of this finding, consider the example of population variability in mutations in the phenylalanine hydroxylase (*PAH*) gene—the gene whose disrupted protein results in the manifestation of phenylketonuria (see above). It has been established that multiple mutations in this gene result in the disorder. The mutations differ in terms of their specific location within the gene, and the frequencies of individual mutations vary across populations. However, each of these mutations appears to arise on one of a limited number of haplotypes and continues to be associated with that haplotype. Most common haplotypes are seen in all populations and the greatest number of haplotypes is seen in African populations.[112]

Third, the essence of the race-intelligence-genetics discussion has been an assumption that if race is somehow a surrogate for unknown genetic mechanisms, then observed racial differences in intelligence and achievement can be explained by genetic differences. But can they be? Although we have gained significant understanding of monogenetic (i.e., single-gene) conditions, there are still enormous blank spots in our understanding of complex human traits (i.e., traits controlled by many genes, often in combination with many environments), such as blood pressure, autism, reading disability, or intelligence. To illustrate, consider the observation that the majority of rare single-gene disorders (e.g., Tay Sachs, sickle-cell anemia, thalassaemia) are caused by mutations in a gene that result in the production of changed and therefore often faulty proteins. In the literature, these deleterious mutations are typically referred to as "coding single nucleotide polymorphisms" (cSNPs). Consider two facts about cSNPs. First, they are rare; second, they are of recent origin, presumably dating to the post-African diaspora.[113] Both assertions have implications for the discussion here.

First, the rarity of cSNPs implies that they are unlikely candidates for controlling quantitative traits such as blood pressure, bone density, and intelligence. The more likely candidates, due to their abundance, are so-called nontranscribed regulatory elements of the genome (i.e., a piece of DNA that does not contribute to the production of proteins, noncoding sequences). The amount of variation in these elements is remarkable. At this point, the significance of this variation, because it has no obvious impact on the proteins, is unclear. However, information from research in other than human organisms is of interest here. For example, in *Drosophila*, these noncoding alleles have been closely associated with quantitative traits.[114] Second, the timing of the origin of cSNPs is linked to the observation that their frequency varies among populations.[115] The reasoning is simple. Because cSNPs arose after the differentiation of the populations, their distribution is a consequence of ethnic differentiation, not a reason for it. It appears that common noncoding variants, some of which are assumed to contribute to or even to underlie susceptibility to common diseases and to variation in quantitative traits, are observed worldwide and can be referred to as "panethnic" alleles.[116] In other words, to the best of our knowledge today, there are no explainable population differences in noncoding allele frequencies that can be meaningfully linked to variation in phenotypes. We simply do not see a clear pattern of ethnic differences in allele frequencies that can be associated with differences in specific phenotypes. Ethnic groups, of course, are socially defined. "Race" sounds like it is biologically defined. It is not. It, too, is socially defined.

SOCIAL VERSUS BIOLOGICAL DEFINITIONS OF RACE

When biological and behavioral markers of socially defined races are investigated, the studies primarily or even exclusively rely on participants' self-reporting of socially defined racial, ethnic, and cultural groups. Many studies use social labels such as Asian American or African American, Chinese, or Hispanic, implicitly ignoring the fact that these labels generalize across substantial amounts of cultural, linguistic, and biological diversity.[117] For example, "Hispanic" includes diverse populations from areas such as Cuba, Puerto Rico, the Dominican Republic, Guatemala, Costa Rica, Argentina, and, of course, Spain. The ancestry of individuals in these groups varies from entirely African, entirely Native American, and entirely European to any possible mixture of these three. Even ignoring the substantial variation within each of these large regions, there is no basis, except for certain social-cultural traits, for grouping these individuals. Even when a more specific populational reference such as Yoruba (i.e., a West-African population of over ten million people who are dispersed throughout different countries in Western Africa) is made, this reference subsumes a great amount of intra-Yoruba variability.[118] Moreover, self-naming of social labels might change, depending on past and present social surroundings of the surveyed participants. For example, during the Soviet era, many immigrating Soviet Jews referred to themselves as Jewish by ethnicity, but upon their arrival to Israel or the United States they referred to themselves as Russians.[119] In the United States, indeed, Judaism is not viewed as an ethnicity, but as a religion. Similarly, individuals who met the classifications of "colored" established by the apartheid government of South Africa would have, probably, self-identified themselves as black in the United States.[120] Thus, because most medical and psychological research on racial differences is based on self-defined racial or ethnic categories and there is substantial evidence questioning the accuracy of these self-classifications, the validity of racial and ethnic differences as commonly investigated is questionable.

People will probably always label themselves and others, regardless of what scientists find. The problem is not the use of social labeling per se, but rather the confusion of it with biological labeling. And it is especially problematical when scientists contribute to this confusion by using social labels in a way that suggests they are somehow biological.

The important message here is that the division lines between racial and ethnic groups "are highly fluid and that most genetic variation exists *within* all social groups—not *between* them.[121] Studies based on hundreds of genetic polymorphisms confirm earlier studies such as that by Lewontin cited above[122] and show that only 11 to 23 percent of observed genetic variation is due to differences among populations and that is mostly attributable to differences in allele frequencies, not all-or-nothing genetic differences.[123] In fact, most common genetic variants exist in almost all populations. The overwhelming majority of the variation occurs among individuals with different genotypes within each population. One study found even less variation among populations, but highly polymorphic multiallelic markers were studied and they may have been biased toward high heterozygosity (i.e., the two chromosomes of an individual having different alleles) in many different populations, thereby minimizing the between-population variation.[124] Clearly, when common polymorphisms are studied, there is only a minority of the genetic variation that occurs among populations. Variants that are restricted to only a few populations in one part of the world are almost never common even in those populations.

Finally, let us regard if and how the concept of race matters in such areas of life as public health and education. Let us consider examples from public health (the data are from the US National Center for Health Statistics, 1998). When age-adjusted death rates of occurrence per 100,000 individuals are reviewed, the rates for white, black, Hispanic, and Asian are as follows: heart disease—121.9, 183.3, 84.2, and 67.4, respectively; cancer—121.0, 161.2, 76.1, and 74.8, respectively; liver disease/cirrhosis—7.1, 8.0, 11.7, and 2.4, respectively; and diabetes mellitus—12.0, 28.8, 18.4, and 8.7, respectively. Three points are important to mention here. First, there are clearly some group differences in these data. However, these differences are inconsistent: for example the incidence of heart disease was the highest among blacks, but the incidence of liver disease was the highest among people of Hispanic origin. Second, all of these conditions are considered to be in part genetic disorders because of the overwhelming amount of data in the field attesting to the importance of the genetic factors in the development and manifestation of these diseases. Third, all these diseases are considered to be complex; therefore, the genetic mechanisms of these conditions have not yet been decoded. Thus, we cannot argue that these observed differences in rates are genetic because we do not know what the genetic mechanisms are.[125]

Similarly, there are some group-average differences in scores on tests of academic abilities and achievement among children socially labeled as white, black, Hispanic, and Asian. How large the differences are, and what groups they favor, depend on what, in particular, is tested. For example, Sternberg and the Rainbow Collaborators found that analytical tests of the kind traditionally used to measure so-called general abilities tend rather strongly to favor Americans of European and Asian origin, but tests of creative and practical thinking show quite different patterns.[126] We also know that there is a substantial genetic influence contributing to individual differences in the level of academic achievement.[127] Yet, we do not know a single gene that has been identified as contributing to either academic achievement or IQ. So, the statement that racial differences in IQ or in academic achievement are of genetic origin is, when all is said and done, a leap of imagination. The literature on intelligence, race, and genetics constitutes, in large part, leaps of imagination to justify, post hoc, social stratifications. There is nothing wrong, in principle, with people expressing their views on social policy. But they need to recognize these views for what they are—social policy pronouncements, not science.

CONCLUSION

In conclusion, the meaning of intelligence is, at this time, ill-defined. Although many investigators study "IQ" or *g* as operational definitions of intelligence, these operationalizations are, at best, incomplete, even according to those who accept the constructs as useful.[128] Research suggests that properties of intelligence beyond *g* may be somewhat different from those of *g*.[129] Race is a social construction, not a biological construct. And studies currently indicating alleged genetic bases of racial differences in intelligence fail to make their point even for these socially defined groups. In general, we need to be careful, in psychological research, to distinguish our folk conceptions of constructs from the constructs themselves.

NOTES

This chapter represents an update to an article originally published by Robert J. Sternberg, Elena L. Grigorenko, and Kenneth K. Kidd in 2005 in the *American Psychologist*. The updates in this chapter were primarily undertaken by Robert J. Sternberg and Steven E. Stemler.

1. R. J. Herrnstein and C. Murray, *The Bell Curve* (New York: Free Press, 1994); J. P. Rushton and A. R. Jensen, "Thirty Years of Research on Race Differences in Cognitive Ability," *Psychology, Public Policy, and Law* 11, no. 2 (2005), 235–94.
2. "'Intelligence and its measurement': A Symposium (1921)," *Journal of Educational Psychology* 12: 123–47, 195–216, 271–75; R. J. Sternberg and D. K. Detterman, eds., *What Is Intelligence?* (Norwood, NJ: Ablex Publishing Corporation, 1986).
3. R. J. Sternberg and E. L. Grigorenko, *Dynamic Testing* (New York: Cambridge University Press, 2002).

4. C. Brand, "Doing Something About *g*," *Intelligence*, 22, no. 3 (1996), 311–26; A. R. Jensen, *The g Factor: The Science of Mental Ability* (Westport, CT: Praeger/Greenwoood, 1998); R. J. Sternberg and E. L. Grigorenko, eds., *The General Factor of Intelligence: How General Is It?* (Mahwah, NJ: Lawrence Erlbaum Associates, 2002).
5. J. B. Carroll, *Human Cognitive Abilities: A Survey of Factor-Analytic Studies* (New York: Cambridge University Press, 1993); J. E. Gustafsson, "Hierarchical Models of Intelligence and Educational Achievement," in *Intelligence, Mind, and Reasoning: Structure and Development*, ed. A. Demetriou and A. Efklides (Amsterdam: North-Holland/Elsevier Science Publishers, 1994), 45–73; J. L. Horn, "Theory of Fluid and Crystallized Intelligence," in *The Encyclopedia of Human Intelligence*, ed. R. J. Sternberg (New York: Macmillan, 1994), 443–51.
6. R. B. Cattell, *Abilities: Their Structure, Growth and Action* (Boston: Houghton Mifflin, 1971); P. E. Vernon, *The Structure of Human Abilities* (London: Methuen, 1971).
7. R. J. Sternberg, *Successful Intelligence* (New York: Plume, 1997); R. J. Sternberg, G. B. Forsythe, J. Hedlund et al., *Practical Intelligence in Everyday Life* (New York: Cambridge University Press, 2000).
8. H. Gardner, "Are There Additional Intelligences? The Case for Naturalist, Spiritual, and Existential Intelligences," in *Education, Information, and Transformation*, ed. J. Kane, (Upper Saddle River, NJ: Prentice-Hall), 111–31; H. Gardner, *Intelligence Reframed: Multiple Intelligences for the 21st Century* (New York: Basic Books, 1999).
9. J. W. Berry, "Radical Cultural Relativism and the Concept of Intelligence," in *Culture and Cognition: Readings in Cross-Cultural Psychology*, ed. J. W. Berry and P. R. Dasen (London: Methuen 1974), 225–29; R. J. Sternberg and J. C. Kaufman, "Human Abilities," *Annual Review of Psychology* 49 (1998): 479–502.
10. The general intelligence factor (abbreviated *g*) is a construct used in the field of psychology to quantify what is common to the scores of all intelligence tests. It was discovered in 1904 by Charles Spearman and subsequently developed in a theory in 1923.
11. R. J. Sternberg, B. E. Conway, J. L. Ketron, et al., "People's Conceptions of Intelligence," *Journal of Personality and Social Psychology* 41 (1981), 37–55; R. J. Sternberg, "Implicit Theories of Intelligence, Creativity, and Wisdom," *Journal of Personality and Social Psychology* 49 (1985), 607–27.
12. Sternberg, "Implicit Theories."
13. E. L. Grigorenko, P. W. Geissler, R. Prince, et al., "The Organization of Luo Conceptions of Intelligence: A Study of Implicit Theories in a Kenyan Village," *International Journal of Behavior Development* 25 (2001): 367–78; R. J. Sternberg, ed., *International Handbook of Intelligence* (New York: Cambridge University Press, 2004); S. Yang and R. J. Sternberg, "Conceptions of Intelligence in Ancient Chinese Philosophy," *Journal of Theoretical and Philosophical Psychology* 17 (1997): 101–19; S. Yang and R. J. Sternberg, "Taiwanese Chinese People's Conceptions of Intelligence," *Intelligence* 25 (1997): 21–36.
14. E. G. Boring, "Intelligence as the Tests Test It," *New Republic*, June 6, 1923, 35–37.

15. N. J. Mackintosh, *IQ and Human Intelligence* (Oxford: Oxford University Press, 1998).
16. R. J. Sternberg,. "Culture and Intelligence," *American Psychologist* 59 (2004): 325–38; R. J. Sternberg, "The Theory of Successful Intelligence as a Basis for New Forms of Ability Testing at the High School, College, and Graduate School Levels," in *Intelligent Testing: Integrating Psychological Theory and Clinical Practice*, ed. J. C. Kaufman (New York: Cambridge University Press, 2009), 113–47.
17. R. J. Sternberg, "Culture, Instruction, and Assessment," *Comparative Education* 43, no. 1 (2007): 5–22.
18. R. J. Sternberg, "Abilities Are Forms of Developing Expertise," *Educational Researcher* 27 (1998): 11–20; R. J. Sternberg, "Intelligence as Developing Expertise," *Contemporary Educational Psychology* 24 (1999): 359–75; R. J. Sternberg, "What is an Expert Student?" *Educational Researcher* 32, no. 8 (2003): 5–9.
19. Cattell, *Abilities*; Horn, "Theory of Fluid and Crystallized Intelligence."
20. Cattell, *Abilities*; Horn, "Theory of Fluid and Crystallized Intelligence."
21. R. B. Cattell and A. K. Cattell, *Test of g: Culture Fair, Scale 3* (Champaign, IL: Institute for Personality and Ability Testing, 1963).
22. J. R. Flynn, "The Mean IQ of Americans: Massive Gains 1932 to 1978," *Psychological Bulletin* 95 (1984): 29–51; J. R. Flynn, "Massive IQ Gains in 14 Nations," *Psychological Bulletin* 101 (1987): 171–91; J. R. Flynn, *What Is Intelligence?* (New York: Cambridge University Press, 2009); U. Neisser, ed., *The Rising Curve* (Washington, DC: American Psychological Association, 1998).
23. S. A. Tishkoff, E. Dietzsch, W. Speed et al. "Global Patterns of Linkage Disequilibruim at the CD4 Locus and Modern Human Origins," *Science* 271 (1996): 1380–87; S. A. Tishkoff, F. A. Reed, F. R. Friedlaender et al., The Genetic Structure and History of Africans and African Americans," *Science* 324 (2009): 1035–44; S. Tishkoff and K. K. Kidd, "Biogeography of Human Populations: Implications for 'Race,'" *Nature Genetics* 36 (2004): S21–S27; R. C. Walter, R. T. Buffler, J. H. Bruggemann et al., "Early Human Occupation of the Red Sea Coast of Eritrea During the Last Interglacial," *Nature* 405 (2000): 65–69.
24. C. B. Stringer, "The Emergence of Modern Humans," *Scientific American*, December 1990, 98–104.
25. Ectodysplasin-A receptor (EDAR) is a cell surface receptor involved in the development of hair follicles, teeth, and sweat glands. EDAR experienced strong positive selection among East Asians.
26. A. Fujimoto, R. Kimura, J. Ohashi et al., "A Scan for Genetic Determinants of Human Hair Morphology: EDAR is Associated with Asian Hair Thickness," *Human Molecular Genetics* 17, no. 6 (2008): 835–43; C. Mou, H. A. Thomason, P. M. Wilan et al., "Enhanced ectodysplasin-A Receptor (EDAR) Signaling Alters Multiple Fiber Characteristics to Produce East Asian Hair Form," *Human Mutation* 29, no. 12 (2008): 1405–11.
27. I. Halder, M. Shriver, M. Thomas et al., "A Panel of Ancestory Informative Markers for Estimating Individual Biogeographical Ancestory and Admixture from Four

Continents: Utility and Applications," *Human Mutation* 29, no. 5 (2008): 648–58; H. Liu, F. Prugnolle, A. Manica et al., "A Geographically Explicit Genetic Model of Worldwide Human-Settlement History," *American Journal of Human Genetics* 79, no. 2 (2006): 230–37; N. A. Rosenberg, J. K. Pritchard, J. L. Weber et al., "Genetic Structure of Human Populations," *Science* 298 (2002): 2381–85.

28. A. R. Templeton, The Genetic and Evolutionary Significance of the Human Race," in *Race and Intelligence: Separating Science from Myth*, ed. J. M. Fish (Mahwah: NJ: Lawrence Erlbaum Associates, 2002), 31–56; J. O. Korbel, P. M. Kim, X. Chen et al. "The Current Excitement About Copy-Number Variation: How It Relates to Gene Duplications and Protein Families," *Current Opinion in Structural Biology* 18 (2008): 366–74.
29. L. M. Cook, "The Rise and Fall of the *Carbonaria* Form of the Peppered Moth," *The Quarterly Review of Biology* 78 (2003): 399–417.
30. R. J. Sternberg, *Psychology*, 4th ed. (Belmont, CA: Wadsworth, 2004).
31. Skin Color Adaptation, http://anthro.palomar.edu/adapt/adapt_4.htm (accessed August 20, 2004.
32. R. M. Harding, E. Healy, A. J. Ray et al., "Evidence for Variable Selective Pressures at MC1R," *American Journal of Human Genetics* 66 (2000): 1351–61.
33. A copy number variation (CNV) is a segment of DNA that has a different number of repeats when comparing two or more genomes. D. F. Conrad, D. Pinto, R. Redon et al., "Origins and Functional Impact of Copy Number Variation in the Human Genome," *Nature*, advance online publication October 7, 2009; E. H. J. Cook and S. W. Scherer, "Copy-Number Variations Associated with Neuropsychiatric Conditions," *Nature* 455 (2008): 919–923.
34. H. Gardner, "Are There Additional Intelligences? The Case for Naturalist, Spiritual, and Existential Intelligences," in *Education, Information, and Transformation*, ed. J. Kane (Upper Saddle River, NJ: Prentice-Hall, 1999), 111–31; H. Gardner, *Intelligence Reframed.*
35. K. K. Kidd, A. K. Pakstis, W. C. Speed et al., "Understanding Human DNA Sequence Variation," *Journal of Heredity* 95 (2004): 406–20; Rosenberg, "Genetic Strucure of Human Populations"; Tishkoff et al., "Genetic Structure"; Tishkoff and Kidd, "Biogeography of Human Populations."
36. K. D. Brownell and K. B. Horgen, *Food Fight: The Inside Story of the Food Industry, America's Obesity Crisis, and What We Can Do About It* (New York: McGraw-Hill, 2003).
37. Kidd et al., "Understanding Human DNA Sequence Variation."
38. J. Marks, "Folk Heredity," in *Race and Intelligence: Separating Science from Myth*, ed. J. M. Fish (Mahwah, NJ: Lawrence Erlbaum Associates, 2002), 95–112.
39. K. Skorecki, S. Selig, S. Blazer et al., "Y chromosomes of Jewish Priests," *Nature* 385 (1997): 32.
40. J. M. Fish, "The Myth of Race," In *Race and Intelligence: Separating Science from Myth.*
41. Ibid.

42. J. P. Rushton, *Race, Evolution, and Man* (Princeton, NJ: Princeton University Press, 1995).
43. R. J. Sternberg, K. Nokes, P. W. Geissler et al. "The Relationship Between Academic and Practical Intelligence: A Case Study in Kenya," *Intelligence* 29 (2001): 401–18; Sternberg, *International Handbook of Intelligence.*
44. R. E. Nisbett and D. Cohen, *Culture of Honor* (Boulder, CO: Westview, 1996).
45. M. C. Campbell and S. A. Tishkoff, "African Genetic Diversity: Implications for Human Demographic History, Modern Human Origins, and Complex Disease Mapping," *Annual Review of Genomics and Human Genetics* 9 (2008): 403–33; F. A. Reed and S. A. Tishkoff, "African Human Diversity, Origins, and Migrations," *Current Opinions in Genetics & Development* 16 (2006): 597–605; Tishkoff et al., "Genetic Stucture"; Tishkoff and Kidd, "Biogeography of Human Populations"; S. A. Tishkoff and S. M. Williams, "Genetic Analysis of African Populations: Human Evolution and Complex Disease," *Nature Review Genetics* 3 (2002): 611–21.
46. Fish, "The Myth of Race."
47. See Herrnstein and Murray, *The Bell Curve*, in their discussion of meritocracy.
48. R. C. Lewontin, "The Apportionment of Human Diversity," in *Critical Race Theory*, ed. E. N. Gates (New York: Garland Publishing, 1997), 7–24.
49. Fish, "The Myth of Race."
50. Ibid.
51. S. J. Gould, "The Geometer of Race," *Discover*, November 1994, 65–69.
52. L. Schiebinger, *Nature's Body* (Boston, MA: Beacon Press, 1993).
53. Gould, "The Geometer of Race."
54. D. Kevles, *In the Name of Eugenics: Genetics and the Uses of Human Heredity* (Cambridge, MA: Harvard University Press, 1995).
55. J. Huxley, "Genetics, Evolution, and Human Destiny," in *Genetics in the Twentieth Century: Essays on the Progress of Genetics During its First 50 Years*, ed. J. L. Dunn (New York: Macmillan, 1951), 501–621.
56. F. Boas, *Race, Language, and Culture* (Chicago, IL: University of Chicago Press), 1942.
57. R. Lynn, *Race Differences in Intelligence: An Evolutionary Analysis* (Augusta, GA: Washington Summit Publishers, 2006); Rushton, *Race, Evolution, and Man*; Rushton and Jensen, "Thirty Year."
58. A. Smedley, *Race in North America: Origin and Evolution of a Worldview* (Boulder, CO: Westview Press, 1993).
59. Lynn, *Race Differences in Intelligence*; R. Lynn, *The Global Bell Curve: Race, IQ, and Inequity Worldwide* (Augusta, GA: Washington Summit Publishers, 2008); Rushton and Jensen, "Thirty Years."
60. Tishkoff et al., "Genetic Structure."
61. D. I. Templer and H. Arikawa, "Temperature, Skin Color, Per Capita Income, and IQ: An International Perspective," *Intelligence* 34 (2006): 21–139.
62. Lynn, *The Global Bell Curve.*
63. N. J. Mackintosh, Book review of *Race Differences in Intelligence: An Evolutionary Hypothesis, Intelligence* 35 (2007): 94–96; R. J. Sternberg and E. Hunt, "Sorry, Wrong

Numbers: An Analysis of a Study of a Correlation Between Skin Color and IQ," *Intelligence* 34 (2006):131–37.

64. W. S. Robinson, "Ecological Correlations and the Behavior of Individuals," *American Sociological Review* 15 (1950): 350–57.
65. J. Hox, *Multilevel Analysis* (Mahwah, NJ: Lawrence Erlbaum Associates, 2002).
66. I. G. G. Kreft and J. de Leeuw, *Introducing Multilevel Modeling* (Thousand Oaks, CA: Sage, 1998); I. G. G. Kreft, J. de Leeuw, and L. Aiken, "The Effect of Different Forms of Centering in Hierarchical Linear Models," *Multivariate Behavioral Research* 30 (1995): 1–22, at 4.
67. T. A. LaVeist, *Race, Ethnicity, and Health: A Public Health Reader* (San Francisco, CA: Jossey-Bass, 2002).
68. R. E. Nisbett, *Intelligence and How to Get It* (New York: Norton, 2009).
69. F. P. Rivara and L. Finberg, "Use of the Terms Race and Ethnicity," *Archives of Pediatrics & Adolescent Medicine* 155 (2001): 119.
70. R. Bhopal and L. Donaldson, "White, European, Western, Caucasian, or What? Inappropriate Labeling in Research on Race, Ethnicity, and Health," *American Journal of Public Health* 88 (1998): 1303–7; M. T. Fullilove, "Abandoning 'Race' as a Variable in Public Health Research: An Idea Whose Time Has Come," *American Journal of Public Health* 88 (1998): 1297–98; R. S. Schwartz, "Racial Profiling in Medical Research," *New England Journal of Medicine* 344 (2001): 1392–93.
71. N. Risch, E. Burchard, E. Ziv, and H. Tang, "Categorization of Humans in Biomedical Research: Genes, Race and Disease," *Genome Biology* 3, no. 7 (2002): 1–12.
72. M. McGue, T. J. Bouchard, W. G. Iacono et al. "Behavioral Genetics of Cognitive Ability: A Life-Span Perspective," in *Nature, Nurture, and Psychology*, ed. R. Plomin and G.E. McClearn (Washington, DC: American Psychological Association, 1993), 59–76; O. S. Davis, C. M. Haworth, and R. Plomin, "Dramatic Increase in Heritability of Cognitive Development from Early to Middle Childhood: An 8-year Longitudinal Study of 8,700 Pairs of Twins," *Psychological Science* 20 (2009): 1301–8.
73. Flynn, *What is Intelligence?*
74. Ibid., 39.
75. Davis et al., "Dramatic Increase in Heritability."
76. N. Risch, E. Burchard, E. Ziv et al. "Categorization of Humans in Biomedical Research: Genes, Race and Disease," *Genome Biology* 3 (2002): 1–12.
77. Risch et al., "Categorization of Humans in Biomedical Research."
78. A. R. Templeton, "Human Races: A Genetic and Evolutionary Perspective," *American Anthropologist* 100 (1999): 632–50.
79. X. Anon, "Genes, Drugs and Race," *Nature Genetics* 29 (2001): 239–40; Schwartz, "Racial Profiling in Medical Research."
80. L. H. Snyder, "Old and New Pathways in Human Genetics," in *Genetics in the Twentieth Century*, 369–92.
81. Tishkoff and Kidd, "Biogeography of Human Populations."
82. *American Heritage Dictionary of the English language*, 4th ed. (Boston: Houghton-Mifflin, 2000), 1441.

83. R. Plomin, "Identifying Genes for Cognitive Abilities and Disabilities," in *Intelligence, Heredity, and Environment*, ed. R. J. Sternberg and E. L. Grigorenko (New York: Cambridge University Press, 1997), 89–104; R. Plomin and F. M. Spinath, "Intelligence: Genetics, Genes, and Genomics," *Journal of Personality and Social Psychology* 86 (2004): 112–29.
84. L. M. Butcher, O. S. P. Davis, L. W. Craig et al., "Genome-wide Quantitative Trait Locus Association Scan of General Cognitive Ability Using Pooled DNA and 500K Single Nucleotide Polymorphism Microarrays," *Genes, Brain and Behavior* 7 (2008): 435–46; S. Buyske, M. E. Bates, N. Gharani, et al. "Cognitive Traits Link to Human Chromosomal Regions," *Behavior Genetics* 36 (2006): 65–76; D. M Dick, F. Aliev, L. Bierut, et al., "Linkage Analyses of IQ in the Collaborative Study on the Genetics of Alcoholism (COGA) Sample," *Behavior Genetics* 36 (2006): 77–86; M. Luciano, M. J. Wright, D. L. Duffy, et al., Genome-wide Scan of IQ Finds Significant Linkage to a Quantitative Trait Locus on 2q," *Behavior Genetics* 36 (2006):45–55; D. Posthuma, M. Luciano, E. J. Geus, et al., "A Genomewide Scan for Intelligence Identifies Quantitative Trait Loci on 2q and 6p," *American Journal of Human Genetics* 77 (2005): 318–26; M. A. Wainwright, M. J. Wright, M. Luciano, et al., "A Linkage Study of Academic Skills Defined by the Queensland Core Skills Test," *Behavior Genetics* 36 (2006): 56–64.
85. S. D. Mandelman and E. L. Grigorenko, Intelligence: Genes, Environments, and Everything in Between," in *The Cambridge Handbook of Intelligence*, ed. R. J. Sternberg and S. Kaufman (New York: Cambridge University Press, 2011.)
86. Plomin and Spinath, "Intelligence: Genetics, Genes, and Genomics."
87. M. J. Chorney, K. Chorney, N. Seese et al., "A Quantitative Trait Locus Associated with Cognitive Ability in Children," *Psychological Science* 9 (1998): 159–66; L. Hill, M. C. Chorney, and R. Plomin, "A Quantitative Trait Locus (Not) Associated with Cognitive Ability?" *Psychological Science* 13 (2002): 561–62; D. M. Dick, F. Aliev, J. Kramer et al., "Association of CHRM2 with IQ: Converging Evidence for a Gene Influencing Intelligence," *Behavioral Genetics* 37 (2007): 265–72; R. Plomin, G. E. McClearn, D. L. Smith et al., "Allelic Associations Between 100 DNA Markers and High Versus Low IQ," *Intelligence* 21 (1995): 31–48; J. R. Zinkstok, O. de Wilde, T. A. van Amelsvoort et al., "Association Between the DTNBP1 Gene and Intelligence: a Case-control study in Young Patients with Schizophrenia and Related Disorders and Unaffected Siblings," *Behavioral and Brain Functions* 3 (2007): 19.
88. R. Plomin, L. Hill, I. Craig et al., "A Genome-wide Scan of 1842 DNA Markers for Allelic Associations with General Cognitive Ability: a Five-stage Design using DNA Pooling and Extreme Selected Groups," *Behavior Genetics* 31 (2001): 497–509.
89. I. J. Deary, W. Johnson, and L. M. Houlihan, "Genetic Foundations of Human Intelligence," *Human Genetics* 126 (2009): 215–32 (2009), at 215.
90. R. J. Sternberg and E. L. Grigorenko, "Myths in Psychology and Education Regarding the Gene Environment Debate," *Teachers College Record* 100 (1999): 536–53.
91. R. Plomin, J. C. DeFries, and G. McClearn, *Behavioral Genetics: A Primer* (New York: Freeman, 1990).

92. N. Block, "How Heritability Misleads About Race," *Cognition* 56 (1995): 99–128 (1995).
93. See Herrnstein and Murray, *The Bell Curve.*
94. T. J. Bouchard, Jr., "IQ Similarity in Twins Reared Apart: Findings and Responses to critics," in *Intelligence, Heredity, and Environment*, ed. R. J. Sternberg and E. L. Grigorenko (New York: Cambridge University Press, 1997), 126–160.
95. R. J. Herrnstein, *IQ in the Meritocracy* (Boston: Atlantic Monthly Press, 1973).
96. R. C. Lewontin, "Annotation: the Analysis of Variance and the Analysis of Causes." *American Journal of Human Genetics* 26 (1974): 400–11.
97. Herrnstein and Murray, *The Bell Curve.*
98. R. C. Lewontin, "The Apportionment of Human Diversity"; R. Lewontin, *Human Diversity* (New York: Freeman, 1982).
99. G. Barbujani, A. Magagni, E. Minch et al., "An Apportionment of Human DNA Diversity," *Proceedings of the National Academy of Science* 94 (1997): 4516–19; Kidd et al., "Understanding Human DNA Sequence Variation"; Rosenberg et al. "Genetic Structure"; Tishkoff and Kidd, "Biogeography of Human Populations."
100. Templeton, "Human Races."
101. Neisser, *The Rising Curve.*
102. R. E. Nisbett, "Race, IQ, and Scientism," in *The Bell Curve Wars*, ed. S. Fraser (New York: Basic Books, 1995); R. E. Nisbett, "Race, Genetics, and IQ," in *The Black-White Test Score Gap*, ed. C. Jencks and M. Phillips (Washington, DC: Brookings Institution, 1998), 86–102; Nisbett, *Intelligence and How to Get It.*
103. S. Scarr and R. A. Weinberg, "IQ Test Performance of Black Children Adopted by White Families," *American Psychologist* 31 (1976): 726–39; S. Scarr and R. A. Weinberg, "The Minnesota Adoption Studies: Genetic Differences and Malleability," *Child Development* 54 (1983): 260–67.
104. M. S. Egorova, "Genotip i sreda v variativnosti kognitivnykh phunktsii [Genotype and Environment in the Variation of Cognitive Functions]," in *Rol' sredy I nasledstvennosti v formirovanii individual'nosti cheloveka*, ed. I. V. Ravich-Shcherbo (Moscow: Pedagogika, 1988), 181–235; E. L. Grigorenko, *Esperimental'nor issledovanie protsessa vydvizheniia I proverki gipotez* [Experimental Study of Hypothesis-Making in the Structure of Cognitive Activity], unpublished doctoral dissertation, NIOPP APN SSSR, 1990; N.V. Iskoldsky, *Vliianie sotsial'no-psikhologicheskikh factorov na individual'nye osobennosti bliznetsov i ikh vnutriparnoe skhodstvo* [The Influence of Social-Psychological Factors Influencing Twins' Individual Characteristics and Their Similarity on Psychological Traits], unpublished doctoral dissertation, NIOPP APN SSSR, 1988.
105. D. Bratko, "Twin Study of Verbal and Spatial Abilities," *Personality and Individual Differences* 21 (1996): 621–24; R. Lynn and K. Hattori, "The Heritability of Intelligence in Japan," *Behavior Genetics* 20 (1990): 545–46; S. Nathwar and P. Puri, "A Comparative Study of MZ and DZ Twins on Level I and Level II Mental Abilities and Personality," *Journal of the Indian Academy of Applied Psychology* 21 (1995): 87–92; S. Pal, R. Shyam, and R. Singh, "Genetic Analysis of General Intelligence 'g': A Twin Study," *Personality and Individual Differences* 22 (1997): 779–80.

106. M. J. Bamshad, S. Wooding, W. S. Watkins et al., "Human Population Genetic Structure and Inference of Group Membership," *American Journal of Human Genetics* 72 (2003): 578–89; M. Bamshad, S. Wooding, B. A. Salisbury et al., "Deconstructing the Relationship Between Genetics and Race," *Nature Reviews Genetics* 5 (2004): 598–609; Kidd et al., "Understanding Human DNA Sequence Variation"; Tishkoff and Kidd, "Biogeography of Human Populations."
107. P. A. Underhill, P. Shen, A. A. Lin et al., "Y Chromosome Sequence Variation and the History of Human Populations," *Nature Genetics* 26 (2000): 358–61; Kidd et al., "Understanding Human DNA Sequence Variation"; L. Quintana-Murci, O. Semino, H. J. Bandelt et al., "Genetic Evidence of an Early Exit of Homo Sapiens Sapiens from Africa Through Eastern Africa," *Nature Genetics* 23 (1999): 437–41.
108. http://alfred.med.yale.edu.
109. A. Chakravarti, "Population Genetics: Making Sense Out of Sequence," *Nature Genetics* 21 (1999): 56–60; M. J. Daly, J. D. Rioux, S. F. Schaffner et al., "High-resolution Haplotype Structure in the Human Genome," *Nature Genetics* 29 (2001): 229–32.
110. L. Frank, "Population Genetics: Estonia Prepares for National DNA Database," *Science* 290 (2000): 31; J. Gulcher and K. Stefannson, "Population Genomics: Laying the Groundwork for Genetic Disease Modeling and Targeting," *Clinical Chemistry and Laboratory Medicine* 36 (1998): 523–27.
111. J. F. Wilson, M. E. Weale, A. C. Smith et al., "Population Genetic Structure of Variable Drug Response," *Nature Genetics* 29 (2001): 265–69.
112. J. R. Kidd, A. J. Pakstis, H. Zhao et al., "Haplotypes and Linkage Disequilibrium at the Phenylalanine Hydroxylase Locus, PAH, in a Global Representation of Populations," *American Journal of Human Genetics* 66 (2000): 1882–99.
113. Tishkoff and Williams, "Genetic Analysis of African Populations."
114. T. F. C. Mackay, "Quantitative Trait Loci in Drosophila," *Nature Review Genetics* 2 (2001): 11–20.
115. Risch et al., "Categorization of Humans in Biomedical Research."
116. R. S. Cooper, "Race, Genes, and Health—New Wine in Old Bottles?" *International Journal of Epidemiology* 32 (2003): 23–25.
117. R. S. Cooper, X. Guo, C. N. Rotimi et al., "Heritability of Angiotensin-converting Enzyme and Angiotensinogen: A Comparison of US Blacks and Nigerians," *Hypertension* 35 (2000): 1141–47.
118. Reich et al. 2001.
119. L. I. Gozman, "The Last Empire: A Divorce in the Family of Nations, in *Psychology in Russia: Past, Present, and Future*, ed. E. L. Grigorenko, P. Ruzgis, and R. J. Sternberg (Commack, NY: Nova Science Publishers, 1997), 395–432.
120. L. Braun, "Race, Ethnicity, and Health," *Perspectives in Biology and Medicine* 45 (2002): 159–74.
121. M. W. Foster and R. R. Sharp, "Race, Ethnicity, and Genomics: Social Classifications as Proxies of Biological Heterogeneity," *Genome Research* 12 (2002): 844–50, at 848.
122. Lewontin, "Annotation"; Lewontin, *Human Diversity*.

123. Reviewed in Tishkoff and Kidd, "Biogeography of Human Populations."
124. Rosenberg et al., "Genetic Structure."
125. Cooper, "Race, Genes, and Health."
126. R. J. Sternberg and the Rainbow Project Collaborators, "The Rainbow Project: Enhancing the SAT Through Assessments of Analytical, Practical, and Creative Skills," *Intelligence* 34 (2006): 321–50.
127. D. Luo, L. A. Thompson, and D. K. Detterman, "Phenotypic and Behavioral Genetic Covariation Between Elemental Cognitive Components and Scholastic Measures," *Behavior Genetics* 33 (2003): 221–46.
128. See, for example, Carroll, *Human Cognitive Abilities*.
129. Sternberg et al., "The Relationship Between Academic and Practical Intelligence."

PART VI

CONTEMPORARY CULTURE, RACE, AND GENETICS

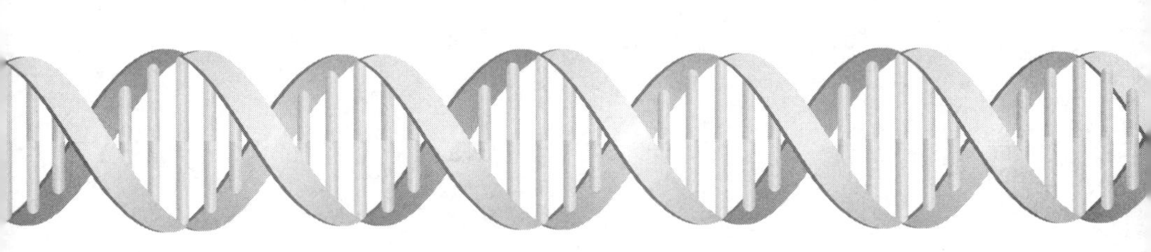

11

THE ELUSIVE VARIABILITY OF RACE

Patricia J. Williams

The question of race is, at its core, a questioning of humanity itself. In various eras and locales, race has been marked by color of skin, texture of hair, dress, musical prowess, digital dexterity, rote memorization, mien, manners, mannerisms, disease, athletic ability, capacity to write poetry, sense of rhythm, sobriety, childlike cheerfulness, animal anger, language, continent of origin, hypodescent, hyperdescent, religious affiliation, thrift, flamboyance, slyness, physical size, contamination, and presence of a moral conscience. As random as such presumed markers may be in the aggregate, they have nevertheless been deployed to rationalize the distribution of resources and rights to some groups and not others. Behind the concept of race, in other words, is a deeper interrogation of what distinguishes beasts from brothers, of who is presumed entitled or dispossessed, person or slave, autonomous or alien, citizen or enemy.

In the contemporary United States, race is based chiefly on broad and variously calibrated metrics of African ancestry.[1] To get a full sense of the ideological incoherence of race and racism, however, one must also include the longer history, in other contexts—whether the centuries-old Chinese condescension to native Taiwanese Islanders,[2] the English derogation of the Irish for "pug noses,"[3] the plight of the Dalit (i.e., untouchables) in India,[4] or comprehensively eugenic regimes like Hitler's, which threw into the ovens Jews, homosexuals, tinkers, conceptual artists, nomadic peoples, the sick, and anyone else designated less than "well-born."[5]

Despite the enormous definitional diversity of what race even means, and though the biological studies—from Charles Darwin's observations to the Human Genome Project—have patiently, repetitively, and definitively shown that all humans are a single species, there remain many determined to reinscribe a multitude of old racialist superstitions onto the biotechnologies of the future. Despite that biological evidence—and in the social sciences, a towering body of social science that is cumulative (observations over time), comprehensive (multiple levels of inquiry), and convergent (from a variety of sources, places, disciplines)—still we are asking the same centuries-old questions.

That said, for purposes of this chapter let us stipulate that race is not a "scientific" or biologically coherent category. I ask for such stipulation because it is beyond my scope to prove or disprove creationist theories of polygenesis, or theological tracts about God's intention to keep races separate, or essentialist polemics about whether black women are more or less endowed with testosterone than white men. It is true that race-as-biology remains a major hurdle in the cultural imagination: at one extreme, there are those zealots who actively deploy races as the innate mark of beings so different that they constitute another species altogether—aliens, sun or moon types, untouchables, nonpersons, beasts. And at the other end of the spectrum are those ordinary creatures for whom discussions of race remain heavily inflected by quiet assumptions of biological difference, within a largely inchoate, unexamined, and unconsciously malleable mush of assumptions about genes, social history, law, and culture.

BIOLOGY AND RACE

So, let us just agree that, as hard as many have tried to find it, there is no allele for race (as distinct from skin color); there are no separate proteins indicating that some of us are chosen by God over others; and there is no distinct cellular pattern that distinguishes the tribal intelligence of any one group on the planet as opposed to another. At the risk of being tedious, I underscore this point precisely because it, like some of the most reproducible of scientific consensuses (evolution, climate change, and the value of vaccinations), remains fiercely disputed as "mere" contestable "theory."

SO WHAT IS RACE IF NOT BIOLOGY?

Race is a hierarchical social construct that assigns human value and group power. Social constructions are human inventions, the products of mind and circumstance. This is not to say that they are imaginary. Racialized taxonomies have real consequences upon biological functions, including the expression of genes. They affect the material conditions of survival—relative respect and privilege, education, wealth or poverty, diet, medical and dental care, birth control, housing options, and degree of stigma, freedom from stigma being something like permission to be happy or to live unburdened by the constant disapprobation of others.

In antebellum America (i.e., the history of the United States before the US Civil War but after the American Revolution and the beginning of the United States as a sovereign nation, from 1789 to 1849), race was determined by a number of variables, depending on the state: color, ancestry, ethnicity, association, behavior, and property records. During the Jim Crow era, a period from 1876 through 1965, state and local laws were enacted in the United States that mandated de jure racial segregation in all public facilities, with a supposedly "separate but equal" status for African Americans. In reality, this led to treatment and accommodations that were inferior to those provided for white Americans, systematizing a number of economic, educational, and social disadvantages. In this period, appearance became foregrounded as singularly important. Since the Civil Rights Movement (1955–68), class and speech have sometimes been included among the criteria of line drawing.

In the industrialized West, racism (as well as related prejudices like class bias, sexism, and religious intolerance) is constructed from a complex intermingling of individual vision, historical happenstance, social milieu, political decision making and legal structure. If not actually rooted in biology, race is nevertheless the subject of relentless biologizing. From the slavery apologist Samuel Cartwright to Adolf Hitler, each generation has brought new utensils to the enterprise of racial demarcation: calipers to measure the size of buttocks or length of leg muscles or circumference of skulls or width of noses.[6] There have been mathematical models to measure percentages of "blood" or wavelengths of skin color or degrees of curvilinearity in the arcs of kinky hair. But over and over, race has been proved and proved again to be illusory as a matter of hard science.

Yet still the questions come: if we are one species, what about subspecies? As in: "Blacks, Jews, Asians—you can't deny they're different. It's like a poodle or a dachshund or a St. Bernard is to the species of dog," according to one of my former students. This sort of perception is not a matter that will be resolved by yet more scientific testing. Rather, I think this reiterated resistance to data is testament to the persistence of human imagination. That we still wonder if there aren't significant disparities in human intelligence that might be logically tracked through the randomness that is race is testament to the power of belief over documentary evidence.

This infernal miasma invites a bit of consideration about the Manichean constructs of determinism and free will, mind and body, choice and constraint, illogic and sheer destiny. Like Dostoevsky's annoying man from the underground we must wonder: Am I a mere piano key, an organ stop? A mathematical inexorability, or a creation of my own intelligent design? The more we tease this out, the more important becomes the narrative lens through which we seek our truths, and the more aware we become of humanity's own constructive power. Am I three-fifths of a human? Ninety-six percent of a chimpanzee? One hundred percent pure tragic mulatta? One-fourth of a nuclear family? An atomistic rational actor? A deficit expenditure of an impoverished underclass?

STUDENTS' RACIAL SELF-IDENTIFICATION

What, in other words, makes "race" both so dangerously essentialized as well as so fleetingly, maddeningly beyond definitional containment? Let's begin with a story. A few years ago, there was an article in the *New York Times* titled "DNA Tells Students They Aren't Who They Thought,"[7] about a sociology class at Pennsylvania State University. Sociology Professor Mark Shriver regularly administers DNA tests to students and has them analyzed for what the article calls "genetic ancestry." Shriver is also a founding partner of the now-defunct company DNAPrint Genomics, which devised a test that "compares DNA with that of four parent populations, western European, west African, east Asian and indigenous American."

The first indication that this was a more romantic than wholly rational enterprise is the classification of these as "parent" populations. The four categories are overly broad for purposes of meaningful ancestry tracking, and unduly and randomly narrow in terms of geographic exclusivity. Given the actual diversity of present-day American populations, the only logic behind this choice of the four groups is that it roughly segregates according to older anthropological descriptions of race-as-color: white, black, yellow, red. And indeed, that's exactly what the students in Shriver's class read into their test results. The article in the *Times* went on for three full columns discussing the degree to which the Penn State students were revealed to be "white" or "black": "About half of the 100 students tested this semester were white," according to an instructor. "And every one of them said, 'Oh man I hope I'm part black,' because it would upset their parents. . . . People want to identify with this pop multiracial culture. They don't want to live next to it, but they want to be part of it. It's cool."[8] But the test purported to show (albeit flawed) geographic origins; it is interesting to see how quickly that was conflated with the matter of color and then from there into the politics of exoticized inclusion against a backdrop of ritual exclusion.

But again, there is no allele for race. As a sociological matter, skin color is a presumptive indicator but historically it is not the exclusive marker. And as a biological matter, melanin concentration merely reveals how one's ancestors adapted to more or less sunny climates—and dark skin is more or less distributed around the equator, no matter which continent. Similarly, evolutionary selection for sickle-cell anemia, often mischaracterized as a "black" disease, is an inherited defensive response to having ancestors who lived among malarial mosquitoes.

That Shriver's test could reveal ancestry based on broad migratory patterns over human history is not a surprise. Certain clusterings of genetic mutations over millennia occur more frequently among specific populations. But those kinship populations cannot be scientifically correlated to the malleable social designations of race.

There is, nevertheless, a remarkable persistence in reinscribing race onto the narrative of biological inheritance. This science is always pursued for only the noblest of reasons: in Shriver's instance, "the potential importance of racial or ethnic background to drug trials."[9] I will save for another paper my concern about the feckless commercial competition for "race-specific" medicines and suggest only that a more coherent enterprise might center on individualized genomic medicine rather than on the ever-changing political variables of racialized bodies.

ANCESTRAL AMALGAMS

For now, consider the description of one student who "discovered" she was "58 percent European and 42 percent African." The young woman "has always thought of herself as half black and half white because her mother is Irish-Lithuanian and her father West Indian."[10] Yet the "parent populations" tested for were described only as "western European" and "west African." Lithuania is generally considered a part of Eastern Europe, and therefore not technically part of the population tested for. While "West Indian" is clearly used as a cipher for her African ancestry, one can be "white"—like Alexander Hamilton—while being West Indian. And the Irish were not considered white in colonial times. Similarly, East Asians have gone in and out of being considered white in our history. South Asians, many being the closest descendants of the original "Aryans," are generally not thought of as white in this country. Yet the incoherent use of Aryan is apparent in any dictionary; according to Merriam-Webster's online edition, the definitions include "Indo-European," "Nordic," and "Gentile."[11]

The degree to which these indivisible habits of thought work despite us, or unintentionally, is perhaps evident in what the Irish-Lithuanian–West Indian student—the one who thought she was half and half—had to say about the test results: "I was surprised at how much European I was, because though my father's family knows there is a great-great-grandfather who was Scottish, no one remembered him . . . I knew it was true, because I have dark relatives with blue eyes, but to bring it up a whole 8 percent, that was shocking to me." What is remarkable—yet not uncommon as a cultural construct—is her flat conception of half and half ancestry, a kind of assumed "purity" of blackness and whiteness. One side had to be entirely African by her measure, one side entirely European. If she's 58 percent European, she assumes the embodied 8 percent must be on the "black" side. The discussion never moves into the more difficult recognition that most West Indians probably have more than 8 percent European ancestry (but, like so many American families, hers might "know" but "not remember" the complicated, often clandestine couplings of the slave trade among Europeans, Africans, and indigenous island peoples). It certainly does not seem to occur to anyone that her white parent might also have an African ancestor.

The jumble of who we are, particularly as residents of the Americas, with its centuries of rapid, recent migrations, is not explored in the *Times* article. The single mention of migratory patterns is misleading: the students whose DNA revealed both African and European ancestors were described as "members of the fastest-growing ethnic grouping in the United States . . . mixed race." But to the extent that a DNA swipe shows "mixing," there is nothing "new" about it; our ancestors have been mixing it up since the first mothers left central Africa; in the long-ago, ancient sense, of course, we are all "African."

Not only do genes not assign race, neither do they have anything to say about the cultural practices we usually refer to as ethnicity or identity. The absurdity of thinking otherwise is highlighted by one of the Penn State students, a warm-brown-colored young man pictured in cornrows, who said that even though he tested at "48 percent European" he values his blackness, since "both my parents are black." He went on to muse, "Just because I found out I'm white, I'm not going to act white." The article ended with an observation that "whatever his genes say," the young man will likely always "be seen as black—at least by white Americans." Consider the narratives therein: Genes "speak" race; whiteness is a biological inheritance that can be consciously "acted"; blackness is defined by the eye of the white beholder.

OBAMA'S "RACE"

If history has shown us anything, it's that race is contradictory and unstable. Yet our linguistically embedded notions of race seem to be on the verge of transposing themselves yet again into a context where genetic percentages act as the ciphers for culture and status, as well as economic and political attributes. In another generation or two, the privileges of whiteness may indeed be extended to those who are "half" this or that. Indeed, some of the discussions about candidate Barack Obama's "biracialism" seemed to invite precisely such an interpretation. Let us not mistake it for anything like progress, however: biracialism always has a short shelf life, and by the time he was elected, President Obama was not our first "half and half president" but had become all African American all the time. Indeed, Obama himself seemed to acknowledge the more complex reality of his own lineage in an off-the-cuff aside, when, speaking about his daughters' search for a puppy, he observed that most shelter dogs are "mutts like me."

In fact, of course, we're all mutts. And as Americans, we've been mixing it up faster and more thoroughly than any place on earth. At the same time, we live in a state of tremendous denial about the rambunctiousness of our recent lineage. The language by which we assign racial category narrows or expands our perception of who is more like whom and tells us who can be considered marriageable or untouchable. The habit of burying the relentlessly polyglot nature of our American identity renders us blind to how intimately we are tied as kin, as family, and as intimates.

In the United States's vexed history of color consciousness, antimiscegenation laws (the last of which were struck down only in 1967) enshrined the notion of hypodescent. Hypodescent is a cultural phenomenon whereby the child of parents who come from differing social classes will be assigned the status of the parent with the lower standing. There are many forms—most parts of the Deep South adhered to it with great rigidity, in what is commonly called the "one drop and you're black" rule. Take for example, *New York Times* editor Anatole Broyard, who denied any relation to his darker-skinned siblings and "passed" for most of his adult life: there were many who expressed shock when it was uncovered that he was "really" black.[12]

Some states, like Louisiana, practiced a more gradated form of hypodescent, indicating hierarchies of status with vocabulary like "mulatto," "quadroon," and "octaroon." And even today, and despite our diasporic, fragmented, postmodern cosmopolitanism, there is a thoughtless or unconscious tendency to preserve these taxonomies, no matter how incoherent. Consider Essie Mae Washington-Williams, the daughter Senator Strom Thurmond had by his family's black maid. She lived her life as a "Negro," then as an "African American," and attended an "all-black" college. But in her seventies, when Thurmond's paternity became publicized, she was suddenly redesignated "biracial." Tiger Woods and Kimora Lee Simmons are alternatively thought of as African American or "biracial," but rarely as "Asian American."

In contrast, many parts of Latin America, like Brazil or Mexico, assign race by the opposite process, hyperdescent. That's when those with any ancestry of the dominant social group, such as European, identify themselves as European or white, when they may also have African or Indian parents. As more Latinos have become citizens of the United States, we have interesting examples of this cultural cognitive dissonance: just think about Beyoncé Knowles and Jennifer Lopez. Phenotypically, they look very, very similar. Yet Knowles is generally referred to as black or African American; Lopez is generally thought of as white (particularly among her Latino fan base) or Latina (among the rest of us), but she is never called black or even biracial. Among Native Americans in the United States there is a combination of both hypo- and hyperdescent, encouraged by the interventionist history of the Bureau of Indian Affairs. Anita Hill, for example, is part Creek, but the narrative about her is entirely about African American origin. And membership in many tribes remains closed to those who have any discernable mixture of African ancestry, but not to those with European ancestry.

WHITE PARENTS CONCEIVE BLACK CHILD

The *New York Post* regularly offers up fascinating tabloid renderings of these contradictions in our culture. When Angelina Jolie adopted her son Pax from Vietnam, the *Post* featured a breathless front-page story, complete with what was described as "a stunning mother-child portrait" of the two.[13] Their faces were aglow with interracial bliss. But the lower half of that day's very same front page was given over to a second, more somber story. Entitled "Baby Bungle: White Folks' Black Child," it trumpeted "a Park Avenue fertility clinic's blunder" that "left a family devastated—after a black baby was born to a Hispanic woman and her white husband."

Long Islanders Nancy and Thomas Andrews had had trouble conceiving after the birth of their first daughter. They employed in vitro fertilization and baby Jessica was born. Jessica is darker skinned than either of the Andrews, a condition their obstetrician initially called an "abnormality." She'll "lighten up," said that good doctor. Subsequent paternity tests showed that Nancy's egg was fertilized by sperm other than Tom's. The couple sued.

If this were the end, the story might simply fall within the growing body of other technological mix-ups resulting in what are sometimes called "wrongful birth" suits, for lost eggs, failed vasectomies, malpractice, broken contract, and so on. There is, after all, a legally recognized expectation that a certain standard of care will be observed in the handling of genetic material. What was distinctive about the Andrews case was that the parents also tried to cite (ultimately without success) *Jessica's* pain and suffering for having to endure life as a black person. The Andrews expressed concern that Jessica "may be subjected to physical and emotional illness as a result of not being the same race as her parents and siblings." They were "distressed" that she is "not even the same race, nationality, color . . . as they are." They described Jessica's conception as a "mishap" so "unimaginable" that they had not told many of their relatives. (Telling the tabloids all about it must have come easier.) "We fear that our daughter will be the object of scorn and ridicule by other children," the couple said, because Jessica has "characteristics more typical of African or African-American descent." So "while we love Baby Jessica as our own, we are reminded of this terrible mistake each and every time we look at her . . . each and every time we appear in public."

One wonders what this construction of affairs will do to Jessica when she is old enough to understand. But here's the really interesting part. When I turned to other media accounts I found a picture of the family—from a 2006 greeting card, no less.[14] And Jessica looks exactly like her mother and elder sister. It is true that Jessica is slightly darker than her mother and that her hair is curlier than her sister's, but all three females are pretty clearly African-descended. As one of my students put it, if anything it is the paleness of the father's skin that marks him as the "different" one.

The picture underscored the embedded cultural oddities of this case, the invisibly shifting boundaries of how we see race, extend intimacy, name "difference." According to *The Post*, Ms. Andrews was "Hispanic" and apparently, by the *Post's* calculations, one Hispanic woman plus one white man must equal "a white couple." The mother is "a light-skinned native of the Dominican Republic," which seemed to indicate that while she may not be "white," she's also not "black."

No matter which of many media accounts I looked at, each narrative implied that if the correct sperm had been used, the Andrews would have been guaranteed a lighter-skinned child. But as most Dominicans trace their heritage to some mixture of African slaves, indigenous islanders, and European settlers, and as dark skin color is a dominant trait, it could be that the true sperm donor is as "white" as Mr. Andrews. But that possibility is exiled from the word boxes that contain this child. Not only was Jessica viewed as being of a race apart from either of her parents; she was even designated a different nationality—this latter most startling for its bloodline configuration of citizenship itself. According to this logic, discrimination is no longer a social problem that implicates all of us and our institutions as unloving or underinclusive. Discrimination becomes destiny, the normative response to biologized "abnormality."

DNA ROOTS

Racial constructions not only oppress by normalizing inequality, they can also make the lie of race seem liberating, attractive, romantic. A small digression to clarify what I mean: a few years ago, there was an interesting convergence of inquiries into the nature of truth. James Frey published his book *A Million Little Pieces*,[15] a wholly fictional account that he proffered as personal memoir. When the fraud was discovered, he defended himself saying that the book was concerned with "emotional truth" rather than literal truth. This triggered deep epistemological soul-searching about whether simple lies can signify, represent, or constitute any kind of figurative truth at all. After a swirl of media confusion, a sound tongue-lashing from Oprah Winfrey seemed to seal up the answer as a resounding "not on my dime."

At the same time that Frey's soap opera was playing itself out, researchers in France were searching for any charred relics at the site where Joan of Arc was said to have been burned at the stake. They wanted to subject any putative remains to DNA testing. Why one would want to do this became something of an issue in the European media: she didn't have children, the site of her martyrdom is in dispute, and the legitimacy of any so-called relic would be highly contested. But the pursuit of "the truth" in so attenuated a context raised questions about the hunger for certainty in the face of such uncertainty. What are the limits of historical insight? How many graves shall we dig up to settle old scores? What are the possibilities of knowing absolutely?

At the same time, there was a similar pursuit unfolding in the American media. Harvard Professor Henry Louis Gates was hosting a series exploring his roots and those of a handful of other prominent African American figures, including comedians Chris Tucker and Whoopi Goldberg, scholar Sarah Lawrence-Lightfoot, and, of course, Oprah Winfrey. It was a fascinating series of TV programs, particularly from the perspective of the discipline of history. It revealed the peculiar difficulties of tracking lines of descent through slavery—the sales of human beings that acknowledged no family ties, the absence of last names, and the absence of first names in some cases—and the necessity of consulting not just census records but also "the master's" property holdings for listings of possible relatives. The reconstruction of family history was like an archeological dig, part intergenerational storytelling, part study of migratory patterns, part recovery of commercial transactions, and part science.

The science du jour is, of course, DNA testing. On the one hand, DNA testing can be quite useful in establishing certain kinds of family relation. (Since the program aired, Gates has set up his own ancestry-tracking company, AfricanDNA.) Gates's own test results showed that he had no relation to Samuel Brady, the white patriarch he'd grown up "knowing" as the man who impregnated his great-great-grandmother. Nothing had prepared him for Brady's *not* being his direct ancestor. Indeed, one of Gates's cousins remained adamant that the test must be wrong. If the test was right, he insisted, there would have to be "two truths": one would be the story he grew up with, the other what the DNA says.

Somewhere in between what the DNA says and what shaped the family account is a gap that is something like a lie. A secret passing from black to white? An act of assimilation or aspiration? A myth to hide some shame, some rape? A change of identity to escape to freedom? Yet I do hesitate to think of it as precisely on the same moral level as the kind of "lie" that James Frey is said to have told in his book. There is something very human about the repetition of family stories until they become epic rather than literal, the burying of family secrets, the lying of ancestors, the reinventions of migrants, the accommodations of raw ambition, and the insulations from terrible shame. This is, I suppose, distantly related to James Frey's addled manipulations; it might also be related to, but of a different order than, the magical thinking of mental patients or character-disordered people or victims of great trauma.

CONCLUSION

There is something so commonplace about the kinds of family mysteries that Gates's inquiries reveal, particularly in the American context. It is part of how many, many of our ancestors, regardless of where they came from, reinvented themselves in the Americas. University of Pittsburgh Law School Professor Jessie Allen describes the "magic" of legal remediation as follows: "What ought to have been prevails over the past." Family stories ritualize the past in a very similar way. It is part of what Professor Robert Pollack, head of Columbia University's Center for the Study of Science and Religion, calls the "eschatology of repair."

If there is value to this kind of "emotional truth," if I can be permitted that term, it is important not to confuse it with the sort of truth that DNA tells us. So while DNA can undoubtedly pinpoint certain aspects of our ancestry through sequencing and matching mitochondrial DNA, it does not make literal sense to say, as Gates did to Oprah Winfrey at one point, "You've got education in your genes." Of course, he was speaking metaphorically at that moment, using the human genome as a metaphor for a pattern of socialization, a family habit, a thirst for knowledge modeled by parents. But at other points in the program as well as in our daily parlance, that metaphoric dimension is applied rather more carelessly, and more dangerously. We have a long history of thinking of identity as genetically based, but again, there is no more an allele for being "white" or "Latina" than there is for "education." These are malleable political designations that expand and contract with time and human circumstance.

It behooves us to be less romantic about what all this DNA swabbing reveals. I worry about the craving to "go back to Africa," to "connect with our Yiddishkeit," or to feel like new doors have been opened if we have an

Asian ancestor. The craving, the connection, the newness of those doors is in our heads, not in our mitochondria. It is a process of superimposing the identities with which we were raised upon the culturally embedded, socially constructed imaginings about "the Other" we could be. The fabulous nature of what is imagined can be liberating and invigorating, but it is a fable. If we read that story into the eternity of our bloodlines, if we biologize our history, we will forever be less than we could be.

NOTES

1. For a history of antebellum litigation about what constituted "whiteness," see Ariella Gross, *What Blood Won't Tell: Racial Identity on Trial in America* (Cambridge, MA: Harvard Press, 2008).
2. Melissa Brown, "On Becoming Chinese," in *Negotiating Ethnicities in China and Taiwan*, ed. Melissa J. Brown (Berkeley: University of California Press, 1996), 37–74.
3. Sander Gilman, *Making the Body Beautiful: A Cultural History of Esthetic Surgery* (Princeton, NJ: Princeton University Press, 1999).
4. Ghanshyam Shah, *Dalit Identity and Politics: Cultural Subordination and the Dalit Challenge* (Los Angeles: Sage Publications, 2001).
5. Sander Gilman, *The Jew's Body* (London: Routledge Press, 1991).
6. For an excellent compendium of such experiments, see Harriet Washington, *Medical Apartheid: The Dark History of Medical Experimentation on Black Americans From Colonial Times to The Present* (New York: Doubleday, 2006).
7. Emma Daly, "DNA Tells Students They Aren't Who They Thought," *New York Times*, April 13, 2005.
8. Ibid.
9. Ibid.
10. Ibid.
11. For a definition of "Aryan," see www.merriam-webster.com/dictionary/aryan (accessed April 29, 2010).
12. Brent Staples, "Anotole Broyard: Back When Skin Color was Destiny—Unless You Passes for White," *Race Matters*, www.racematters.org/anotolebroyard.htm (accessed April 29, 2010).
13. Tod Venezia, "Baby Bungle: White Folks' Black Child," *New York Post*, March 22, 2007.
14. Julian Kesner and Nicole Lyn Pesce, "In Vitro Fertilization Risks in the Spotlight," *New York Daily News*, September 23, 2009.
15. James Frey, *A Million Little Pieces* (New York: Doubleday, 2003).

12

RACE, GENETICS, AND THE REGULATORY NEED FOR RACE IMPACT ASSESSMENTS

Osagie K. Obasogie

Despite significant advances in race relations and the status of people of color, racial minorities face new challenges in the twenty-first century that are unmistakably connected to past injustices. An emerging concern is how human biotechnologies are being used to lend support to framing racial disparities and differences as distinctly biological rather than social phenomena. Previously discredited beliefs that inherent biological differences give rise to racial disparities in health and other social outcomes are under increasing reconsideration. In a nutshell, the color line that has and continues to divide racial groups is increasingly taking on, in the view of some, a genetic character.

But these new articulations of biological race have a different overtone from their predecessors. In the name of resolving racial disparities in health, shedding light on disrupted genealogies, and improving law enforcement, they explicitly reject the racial subordination that fueled past efforts to link social categories of race to inherent biological differences. Yet they may inadvertently lead to similar conclusions; various racial disparities—from why certain groups are sicker than others to why arrest and incarceration rates are higher among some populations—may come to be more meaningfully understood through genetic rather than social or environmental mechanisms.

It is important to note from the outset that there is some evidence that social categories of race may be genetically relevant to the extent that they may loosely correlate with geographical origin, broadly defined. This, in turn, may reflect the histories of isolation and evolution experienced by some groups. Yet there is also evidence that many biotech applications that use social categories of race treat them in a circular fashion; the presumption that social categories of race are biologically salient can shape research questions and agendas, as well as data collection and analysis. The preponderance of evidence shows that the distinct racial categories that society has created do not directly align with meaningful population differences. Nevertheless, this growing body of work concerning race and genetics has led to three key applications that may reinvent biological understandings of racial difference and disparity and adversely impact communities of color: race-based medicines, genetic ancestry tests, and DNA forensics.

While each of these applications has been examined individually, it is useful to consider them together to highlight a fundamental concern: that human biotechnology may give new life to the discredited idea that social categories of race reflect discrete biological differences between racial groups. In other words, it is important to ask: are twenty-first-century technologies reinventing nineteenth-century theories of racial difference? To examine this question, it is useful to explore developments in each of these three areas.

RACE-BASED MEDICINES

The Food and Drug Administration's (FDA) June 2005 approval of BiDil as a treatment for African Americans with heart failure was significant in that it marked the first time that regulatory approval had ever been given to a drug with a race-specific indication. Initially marketed by NitroMed as a way to address what were perceived as racial disparities in heart failure, BiDil quickly became the poster child of revamped efforts to approach race not merely as a social category, but as a biologically relevant mechanism for understanding human difference and health outcomes.

BiDil's approval represents at least three different claims about the relevance of race to health care and health disparities. It was (1) the first drug to be patented as race-specific (a *legal* claim about race and biology), (2) the first to receive FDA approval as race-specific (a *regulatory* claim about race and biology), and (3) the first to be marketed as race-specific (an *economic* claim about race and biology). Although no genetic data were presented to the FDA, BiDil's FDA approval marked an important step in government giving legitimacy to framing racial difference as a proxy for significant genetic differences in human populations. Steven Nissen, chair of the FDA's Cardiovascular and Renal Drugs Advisory Committee that endorsed BiDil's approval, could not have been clearer in affirming this, noting that his committee took self-identified race in the clinical trial supporting BiDil's approval (A-HeFT) "as a surrogate for genomic-based medicine."[1]

Many ask, why not support BiDil, if it really helps African Americans who suffer from heart failure? The issue is that much of the evidence supporting this claim is not as convincing as it initially seems. For example, much of the moral impetus behind BiDil's approval was the frequently cited claim that there is a 2:1 racial disparity in heart failure mortality between blacks and whites. But blacks are not twice as likely to die from heart failure as anyone else. Legal scholar Jonathan Kahn, who followed the BiDil story very closely, traces this claim to a series of misquotes concerning what is now quarter-century-old data.[2] More recent data from the Centers for Disease Control and Prevention puts the ratio at 1.1:1. Essentially, there is little to no difference in population-wide mortality between blacks and whites.

Another concern is that the clinical trial used to support the claim that BiDil is a race-specific drug had significant flaws. The A-HeFT trial that propelled BiDil's FDA approval does not clearly support the claims of race specificity made by the drug's proponents. Any clinical trial that yields a 43 percent reduction in mortality is a stunning feat. Yet by only enrolling self-identified blacks, the trial strongly implies (and is indeed used to show) that it is *only* effective in African American populations. But this is simply not the case. Even Dr. Jay Cohn, the person who developed BiDil, acknowledges that nonblacks can receive a substantial benefit from the medication.[3] As Jonathan Kahn notes, "The only responsible scientific claim that can be made on the basis of these trials is that BiDil works in some people who have heart failure, period."[4]

Lastly, there is no evidence that race is a suitable proxy for genetic differences in drug response. While Dr. Nissen and other BiDil supporters argue that self-identified race can be used as a proxy for genetic differences until specific genetic variations are located, no genetic component to BiDil's ostensible race-specific efficacy has been found. Race-based medicine is based, in part, upon the idea that specific genetic variations that are most common within particular populations explain certain health disparities and that these disparities can be remedied with therapies that take such knowledge into consideration. BiDil's clinical trials arguably put the cart before the horse, replacing a scientific approach with the assumption that *perceived* racial difference equals genetic difference connected to heart failure.

The assumptions and missteps embedded in efforts to develop and market race-specific medicines raise several concerns. They contribute to three possible outcomes that may work against sensible approaches to addressing health disparities. First, social determinants of health might take a backseat. Scientific studies that root health disparities in genetic differences could obscure the social and environmental factors that affect groups' disparate health outcomes. Thinking about racial disparities in genetic terms deemphasizes how groups' poor treatment can lead to their poor outcomes. Second, claims about a genetic basis for racial disparities in health outcomes can quickly influence how we understand other social disparities. A key concern is the temptation to use the idea that racial disparities in health reflect inherent genetic differences to explain racial disparities in other areas such as employment, education, and criminal justice. These disparate outcomes might then be attributed to people's genes rather than barriers and privileges connected to social, economic, and political factors and access to resources. Lastly, race-specific medicines can shift the responsibility for resolving racial disparities in health from public health initiatives to private biomedical ventures. This is not to say that profit interests can never converge with genuine opportunities to reduce health disparities. But ceding the problem of racial disparities in health to biomedical companies might devalue public health mechanisms that can tackle these disparities' core social and environmental causes.

Although sales of BiDil have struggled significantly,[5] more race-based medicines may very well be around the corner. University College London biologists Sarah Tate and David Goldstein noted in a 2004 *Nature*

Genetics article that while controversial, "at least 29 medicines (or combination of medicines) have been claimed, in peer reviewed scientific or medical journals, to have differences in either safety or, more commonly, efficacy among racial or ethnic groups."[6] The Pharmaceutical Research and Manufacturers of America (PhRMA), the industry's trade group, released a report in December 2007 noting that its member companies "are developing 691 medicines for diseases that disproportionately affect African Americans or diseases that are among the top 10 causes of death for African Americans . . . [to] help close the health disparity."[7] While this report does *not* specifically pertain to medicines claiming to be genetically tailored for blacks, the report's framing contributes to a perspective that drug companies are promoting and that may become increasingly popular within the biomedical sciences: that health disparities are linked to group predispositions and susceptibilities that are best addressed through targeted medications.

GENETIC ANCESTRY TESTS

Genetic testing is often presented as a major breakthrough in healthcare, as DNA technologies may provide insight into individuals' predisposition for disease and the optimal use of certain drugs. A more questionable approach to these technologies is what some have termed "recreational genetics":[8] DNA tests focused not on health but on giving customers some type of ancillary information, such as insights into their genealogy. The marketing and sale of direct-to-consumer genetic ancestry tests is projected to become a growing industry over the next several years.

One sector is particularly booming: African Americans using genetic tests to discover their ancestral origins, often in an attempt to make an end-run around the genealogical dead end produced by the slave trade. Genetic ancestry tests examine individuals' DNA to see if they have genetic markers similar to those from populations found elsewhere in the world. Some genetic markers are found more frequently in certain populations than others, which may give clues to the geographical origin of an individual's particular genetic sequence.

The technological developments underlying the commercial viability of genetic ancestry tests stem in large part from population genetics, an academic field that looks at how evolutionary forces shape groups' genetic makeup. But there is an often-unnoticed leap of logic between discussions of *group* genetic differences and genetic ancestry tests' ability to reliably say anything meaningful about *individual* ancestry. Group-based studies investigate frequency distributions of different populations' genetic variations whose boundaries are recognized as being inherently blurry; their applicability to the genealogy of any individual is limited. Moreover, there has been insufficient discussion on how translating academic research on groups and populations into commercial ventures on individual ancestry can breathe new life into biological notions of race.

Advocates of genetic genealogy tests rarely use the term "race," preferring terms such as "biogeographical ancestry" or "continental ancestry."[9] But from both scientific and consumer perspectives, genetic ancestry tests raise a series of important issues. Key among these are their likely social outcomes that (1) industry euphemisms such as "biogeographical ancestry" will more often than not be understood as "race" and that (2) the dissemination of the idea that social categories of race can somehow, even minimally, be genetically verified by a simple cheek swab.

Currently, genetic ancestry tests take three main approaches.

1. *Mitochondrial DNA* (*mtDNA*) tests rely on the fact that this specialized part of our DNA is passed only from mother to child (unlike most DNA, which is a mixture from both parents). It can therefore be used to test a direct maternal line.
2. *Y chromosome* tests analyze genetic markers passed from father to son to trace paternal ancestry.
3. *Admixture mapping* examines genetic markers on nonsex chromosomes that contain DNA from both parents to estimate a person's percentages of African, Native American, European, and East Asian ancestry. Significant methodological questions remain concerning these tests' accuracy.

Mitochondrial DNA and Y Chromosome tests can identify whether any two *individuals* are related with a high degree of certainty. However, they are also used to determine which genetic markers an individual might share with a population as a way to give customers a sense of their geographical origins as a proxy for what race they might be. This second use of mtDNA and Y chromosome tests has severe limitations. The reason they are so limited is because both of these tests examine only a very small fraction of the genetic material contributing to an individual's genome. Each of our parents, grandparents, great-grandparents, etc., contribute to our genetic makeup. Going back seven generations, that is 128 great-great-great-great-great-grandparents that have an equal "say" in an individual's genome. Yet taken together, mtDNA and Y chromosome testing *only provide information about two of those ancestors* whose genetic information has been passed down throughout time, presumably unchanged. This raises the question: what about the other 126 contributors? These two tests discount the significance of these 126 contributors not because they are less important, but simply because these tests cannot access their information.

The third type of test, *admixture mapping*, is thought to resolve some of these problems. It checks 175 autosomal markers (autosomes being the 44 nonsex chromosomes inherited from both parents) that are thought to be related to certain ancestral backgrounds. The alleles, or genetic variants, used as markers with admixture mapping are "those that have the most uniqueness, or the largest differences in allele frequency among populations."[10] For example, after sampling populations from around the world, a database might show that one genetic marker is prevalent among samples from West Africans but not Native Americans, leading the admixture test to conclude that any person with this marker has some West African heritage. While most genetic markers do not reflect this type of variation, admixture mapping relies upon the few markers that do and are also connected to a geographically distinct population.[11] This blend of genetic information is thought to be able to convey a better sense of overall ancestry, but admixture mapping has its own limitations. For example, to talk about genes and ancestry in terms of percentages and mixtures seems to presume that racial purity exists, or existed at one time. This can give a misleading impression that genetically distinct populations are real (or were at some point) and that social categories of race are genetically verifiable.[12]

Genetic ancestry tests raise at least three concerns. First, there are no known genetic variations that are exclusive to any socially defined racial group. While researchers may be able to determine that certain genetic variations occur more or less frequently in certain geographically defined populations, they have not shown that these variations align discretely with social categories of race that are largely defined by phenotype or other cultural norms. Second, there are significant limitations with the databases that these companies use. Inferences linking an individual's genetic background to a particular group of people are only as good as the underlying group samples used by genetic ancestry companies. The entire enterprise depends upon data from very small samples of people; what might appear to be clear markers of a certain group's ancestry may, after broader sampling, turn out not to define the group after all.[13] This is the likely reason why individuals who take genetic tests from multiple companies often receive conflicting results about their ancestral backgrounds.[14] Lastly, the claims made by these endeavors often go beyond the current state of the science. Genetic ancestry companies often make statements about their products' precision that are not scientifically supported; consumers can be misled about these tests' ability to accurately pinpoint their ancestral origins. Furthermore, these companies have proprietary interests in their tests and do not disclose the methods used to determine test results. Without more transparency, it is difficult to assess the scientific basis for their conclusions.

Using today's social categories of race and geographical distribution of populations as transcendent reference points from which to understand groups' past identities and locations is not only questionable science, but it also directly contradicts what we do know about the fluidity of social categorizations and migration patterns.[15] At best, using group-based population studies to speak to individuals' ancestral pasts provides a sliver of information about a person's ancestry. At worst, however, these commercial endeavors can give new legitimacy to racial typologies and revive discredited beliefs that social categories of race reflect fixed inherent differences.

DNA FORENSICS

DNA forensics is an important tool that has been used to identify perpetrators and to exonerate people previously found guilty on less reliable evidence. Hundreds of wrongful convictions have been overturned, including scores of people on death row.[16] However, a number of significant questions about DNA forensics are beginning to emerge. It is important to note that the racial justice issues concerning DNA forensics are slightly different than those regarding race-based medicines and genetic ancestry tests. While aspects of DNA forensics leverage social categories of race to develop applications for identifying perpetrators[17] or understanding individual or group propensity for criminality,[18] it raises additional concerns over how new genetic technologies intersect with entrenched racial bias in the criminal justice system.

No one seriously doubts the reliability of DNA typing. Yet, a separate question involves the handling and interpretation of the underlying evidence. Among the issues:

Contamination—If a sample is mixed with other DNA (which can happen at any stage in collection, handling, and testing), both false positives and false negatives can result.

Clerical errors—Opportunities for introducing error arise during the procedures involved with logging samples and computer data entry.

Misinterpretation—When samples are small or old, they are particularly susceptible to being misinterpreted by laboratory personnel. Misinterpretation can also occur in cases of "mixtures," for example, when the DNA from the crime scene consists of a mixture from two or more individuals.

False matches—Random false matches do occur, most likely with close relatives.[19]

In addition to these issues, massive backlogs exist at forensic labs throughout the country. Employees are often under pressure to reduce these backlogs, which can create the conditions for more human error.

Another issue concerns the expanding use of DNA databases, which are at the heart of emerging controversies concerning DNA forensic technologies. These databases store the genetic profiles of felons and, in some jurisdictions, even people arrested or detained for felonies without ever being charged. Police argue that larger collections of genetic profiles will allow rapid identification of offenders who leave behind samples containing DNA and help solve cold or future cases. The rapid expansion of state and federal DNA databases has given rise to a new type of case in law enforcement: the cold hit, where the only evidence linking a suspect to a crime is that biological material left at a crime scene matches a

DNA profile in a database. As DNA databases grow, so too has the cold hit approach to solving crimes. While some take cold hits as unassailable proof of suspects' guilt, much closer scrutiny is warranted.

It is important to distinguish between what forensic scientists call full and partial matches. A full match occurs when crime scene evidence matches a known sample in a database across the thirteen loci standard introduced by CODIS (Combined DNA Index System, the FBI's national database). Matches across fewer than thirteen loci are known as partial matches.[20] Increasingly, partial matches are used even in cold hit cases as incriminating evidence. But new questions about partial matches are emerging. Bicka Barlow, a California attorney representing a defendant implicated in a rape/murder by a cold hit matching across thirteen loci, heard that Arizona's DNA database had two profiles that matched across nine loci. After filing a subpoena to find out more about this, she received a puzzling report: out of 65,493 offenders in Arizona's database in 2005, 122 pairs of people had genetic profiles matching at nine loci, 20 pairs matched at ten loci, one pair of siblings matched at eleven loci, and another pair of siblings matched at twelve.[21] Such findings seem implausible, given the accepted statistical norm that the odds of a random match happening between any two people across nine loci are one in several million. But therein lies the problem: cold-hit matches that occur *within databases* do not reflect the same odds as finding a match *within entire populations.*[22] A one in twenty million probability match to a cold hit in a large forensic database, for example, does not mean that the chances that the profile is not the suspect's is one in twenty million. One way to understand this is through what statisticians call "the birthday problem": although the probability that any one person has a particular birthday is 1 in 365, there is a 50 percent chance that two people will share a particular birthday in a group of twenty-three or more people.[23] While not a perfect parallel to DNA databases, the birthday problem illustrates the often-radically different chances of finding a match when probabilities are expressed in relation to the general population as opposed to a defined number of profiles in a database.

Familial searches within DNA databases are another emerging issue. This entails running an unknown DNA sample across a database to find partial matches that might not directly identify a particular suspect, but may be close enough to point to a suspect's family member who may be responsible for the crime. Using partial matches to identify potential suspects radically expands the power and purpose of DNA databases from the individual to the family, implicating a number of people who may have nothing to do with the original crime. Given these databases' disproportionate composition—for example, it has been estimated that Blacks constitute 40 percent of the federal DNA database while only being 13 percent of the population—racial minorities are the most likely to be implicated in crimes they may very well not have committed.

Another development is the budding practice of molecular photofitting: "methods to produce forensically (or biomedically) useful predictions of physical features or phenotypes from an analysis of DNA variations . . . [to provide] a summary list of physical traits like height, weight, hair color, eye color, and race, and a fuzzy or low resolution picture."[24] Put differently, researchers are working on being able to produce a physical description and picture of a suspect simply by analyzing biological material left at a crime scene.

It is one thing when conflating social categories of race with genetic categories leads to less than accurate understandings of an individual's ancestry. It is quite another when these less than precise mechanisms become part of a criminal justice system where individuals' freedoms are at stake. While law enforcement uses all types of methods to produce leads, the presumed infallibility of DNA technologies can lead prosecutors, judges, juries, and others involved in the criminal justice system to think differently about the evidence in relation to the suspect. Further exacerbating the problem, many if not most defense attorneys lack expertise in forensic DNA analysis. This might lead them to not challenge prosecutorial assertions as vigorously as they might otherwise.

THE NEED FOR RACE IMPACT ASSESSMENTS

Racially tailored medicines, new ways of investigating individual ancestry, and expanding forensic tools for law enforcement are laudable attempts at harnessing the power of biotechnology to improve everyday life. But these and other developments also have the potential to negatively affect communities of color and, moreover, to distort public understandings of race.

We are at a critical moment. Whether new human biotechnologies turn out to disproportionately burden racial minorities and warp lay understandings of race depends heavily upon the care with which researchers, biotech companies, and policymakers treat race in their work. It is crucial that we require sound evidence for any claims attempting to link social categories of race to biological difference. Responsible regulation and oversight can go a long way toward ensuring that these products and services are based on sound scientific research, and that they do not promote unfounded biological theories of racial difference. How can this be accomplished?

To encourage more forethought in regulatory decision making and implementation, other fields have adopted impact assessments. One relevant example is the *health impact assessment*, which is a set of procedures, methods, and tools that can provide a framework for policymakers to predict, map, and mitigate adverse consequences stemming from a policy proposal.[25] For example, a health impact assessment of a proposal for a new factory would look at a number of ways it may affect the local population's health, such as whether emissions from the building are linked to adverse health outcomes and how best to contain them.

Similar regulatory assessments of the possible public impact of an innovation or initiative may be instructive for identifying and mitigating their possible adverse effects for racial minorities. *Race impact assessments* could encourage shared responsibility among multiple actors—such as regulators, researchers, institutional review boards, and affected communities and their representatives—in making sure that human biotechnologies are not used to promote unfounded biological understandings of race and that claims made about the relationship between race and genetics are legitimate.[26] Just as health impact assessments aim "to enhance recognition of societal determinants of health and of intersectoral responsibility for health,"[27] race impact assessments could promote recognition of the social construction of race and the social determinants of racial disparities.

What might such race impact assessments look like in the context of human biotechnology? As an example, legislators and regulators might rethink the Food and Drug Administration's traditional scope of safety and efficacy to convene expert committees that evaluate whether medicines like BiDil might reinforce biological understandings of race when no biological or genetic mechanism has been identified. The composition of such a committee would have to accurately reflect the impacted stakeholders and constituents. Its assessment would not be limited to reviewing biostatistical evidence from clinical trials. It would also consider the effects race-specific medicines might have on broader commitments to racial justice, specifically in the context of past discrimination based on biological notions of race. This might encourage narrowly tailored mechanisms to ensure that a drug's beneficiaries have access without prematurely giving legitimacy to biological understandings of racial difference.

A race impact assessment of ancestry tests might lead federal or state governments to closely scrutinize marketing claims to ensure that they do not overstate the current state of the science. Such assessments might lead regulators to require genetic testing companies to limit their advertising to scientifically verifiable statements, and to give consumers adequate information about the tests' limitations. In the context of DNA forensics, a race impact assessment could shed light on policy shifts that might disproportionately affect certain communities, such as familial searching or including arrestees that have not been convicted in DNA databases. This assessment might encourage refinements and recalibrations that could lessen the burden on those communities while ensuring that law enforcement has the tools it needs.

The overall goal of race impact assessments would be the same as its counterparts in public health and other realms: to increase dialogue between stakeholders and policymakers so as to balance competing interests through strategic planning that promotes the public good. Race impact assessments have the potential to play a key role in ensuring that human biotechnologies develop in a manner that benefits society without unduly burdening racial minorities.

NOTES

This chapter is excerpted from Osagie K. Obasogie, *Playing the Gene Card? A Report on Race and Human Biotechnology*, Center for Genetics and Society (2009). The full report is available at www.thegenecard.org.

1. "BiDil African-American Subset Is Surrogate For Genomics, Cmte. Chair Says," *The Pink Sheet* 67, no. 25 (June 20, 2005): 3.

2. See Jonathan Kahn, "Getting the Numbers Right: Statistical Mischief and Racial Profiling in Heart Failure Research," *Perspectives in Biology and Medicine* 46, no. 4 (autumn 2003): 475–78.
3. "Cohn . . . says he believes it probably will be effective in patients who aren't black. In fact, he says, he prescribes the generic drugs that make up BiDil for the 25 percent of his white patients who don't do well on other drugs. 'I actually think everybody should be using it,' Cohn said." Denise Gellene, "Heart Pill Intended Only for Blacks Sparks Debate," *Los Angeles Times*, June 16, 2005, http://articles.latimes.com/2005/jun/16/business/fi-bidil16.
4. Jonathan Kahn, "From Difference to Disparity: How Race Specific Medicines May Undermine Policies To Address Inequalities in Health Care," *Southern California Interdisciplinary Law Journal* 15 (2006): 106.
5. See Tara Bannow, "Race-related Controversy Causes Drug Flop," *Minnesota Daily*, March 9, 2010, www.mndaily.com/2010/03/09/race-related-controversy-causes-drug-flop.
6. Sarah K. Tate and David B. Goldstein, "Will Tomorrow's Medicines Work for Everyone?" *Nature Genetics Supplement* 36 (November 2004): S34. They go on to note that "these claims are universally controversial and there is no consensus on how important race or ethnicity is in determining drug response."
7. PhRMA, "Nearly 700 Medicines in the Pipeline Offer Hope for Closing the Health Gap for African Americans," December 2007, www.phrma.org/news_room/press_releases/new_report_offers_hope_for_african_americans,_shows_nearly_700_medicines_in_pipeline/ (accessed June 25, 2008).
8. D. Bolnick, D. Fullwiley, T. Duster et al, "Science and Business of Genetic Ancestry Testing," *Science* 318 (2007): 399–400.
9. See, for example, the Genetic Identity Ancestry Testing webpage, www.genetic-identity.com/Ancestry_Testing/ancestry_testing.html (accessed Mar. 25, 2008).
10. Tony N. Frudakis, *Molecular Photofitting: Predicting Ancestry and Phenotype Using DNA* (Burlington MA: Elsevier, 2008), 44
11. Deborah Bolnick, "Individual Ancestry Inference and the Reification of Race as a Biological Phenomenon," in *Revisiting Race in a Genomic Age*, ed. B. Koenig et al. (New Brunswick, NJ: Rutgers University Press, 2008).
12. Deborah A. Bolnick et al., "The Science and Business of Genetic Ancestry Testing," *Science* 318 (October 19, 2007): 399.
13. "For example, genetic ancestry testing can identify some of the groups and locations around the world where a test-taker's haplotypes or autosomal markers are found, but it is unlikely to identify all of them. Such inferences depend on the samples in a company's database, and even databases with 10,000 to 20,000 samples may fail to capture the full array of human genetic diversity in a particular population or region." Deborah A. Bolnick et al., "The Science and Business of Genetic Ancestry Testing," *Science* 318 (October 19, 2007): 399.
14. Ron Nixon, "DNA Tests Find Branches But Few Roots," *New York Times*, November 25, 2007, www.nytimes.com/2007/11/25/business/25dna.html (accessed March 25, 2008).

15. "Consumers often purchase these tests to learn about their race or ethnicity, but there is no clear-cut connection between an individual's DNA and his or her racial or ethnic affiliation. Worldwide patterns of human genetic diversity are weakly correlated with racial and ethnic categories because both are partially correlated with geography." Deborah A. Bolnick et al., "The Science and Business of Genetic Ancestry Testing," 399, 400.
16. Innocence Project, Special Report: *200 Exonerated, Too Many Wrongfully Convicted*, www.innocenceproject.org/Images/751/ip_200.pdf.
17. Frudakis, *Molecular Photofitting*, 16.
18. See Special Issue on the Impact of Behavioral Genetics on the Criminal Law, Law, and Contemporary Problems, 69 (2006): 1–2 (Nina A. Farahany and James E. Coleman, Jr., were the editors of the issue.)
19. See, for example, Kristina Staley, "The Police National DNA Database: Balancing Crime Detection, Human Rights and Privacy," Genewatch UK 22 (January 2005), www.genewatch.org/pub-492774 (accessed Mar. 25, 2008); Tania Simoncelli, "Retreating Justice: Proposed Expansion of Federal DNA Database Threatens Civil Liberties," *GeneWatch* 17 (April 2004): 3, www.gene-watch.org/genewatch/articles/17-2Simoncelli.html (accessed March 25, 2008).
20. "A partial match at 9 loci, for instance, would be a pair of individual who match at 9 CODIS loci out of the 13." Laurence D. Mueller, "Can Simple Population Genetic Models Reconcile Partial Match Frequencies Observed in Large Forensic Databases?" *Journal of Genetics* 87 (July 8, 2008): 2, 100–8.
21. Jon Jefferson, "Cold Hits Meet Cold Facts: Are DNA Matches Infallible?" *Transcript* 40 (2008): 29–33. Barlow's efforts to shed light on this issue have been repeatedly hindered. See, generally, A. C. Thompson, *Weird Science: Why is S. F.'s Crime Lab Resisting Scrutiny by Defense Attorneys?* See www.sfbg.com/39/27/news_crime_lab.html; accessed April 26, 2011.
22. UC Irvine's Bill Thompson explains: "The risk of obtaining a match by coincidence is far higher when authorities search through thousands or millions of profiles for a match than when they compare the evidentiary profile to the profile of a single individual who has been identified as a suspect for other reasons. As an illustration, suppose that a partial DNA profile from a crime scene occurs with a frequency of 1 in 10 million in the general population. If this profile is compared to a single innocent suspect, the probability of a coincidental match is only 1 in 10 million. Consequently, if one finds such a match in a single-suspect case it seems safe to assume the match was no coincidence. By contrast, when searching through a database as large as the FBI's National DNA Index System (NDIS), which reportedly contains nearly 6 million profiles, there are literally millions of opportunities to find a match by coincidence. Even if everyone in the database is innocent, there is a substantial probability that one (or more) will have the 1-in-10 million profile. Hence, a match obtained in a database search might very well be coincidental." William C. Thompson, "The Potential for Error in Forensic DNA Testing (and How That Complicates the Use of DNA Databases for Criminal Investigation)," www.councilforresponsiblegenetics.org/pageDocuments/H4T5EOYUZI.pdf.

23. Karen Norrgard, "Forensics, DNA Fingerprinting, and CODIS," *Nature Education* 1, no. 1 (2008), www.nature.com/scitable/topicpage/Forensics-DNA-Fingerprinting-and-CODIS-736.
24. Frudakis, *Molecular Photofitting*, 16.
25. While an examination of health impact assessments is most relevant for the purposes of this discussion, it is important to acknowledge that health impact assessments have "much in common with and builds on "environmental impact assessment" and also has less recognized but salient links with the field of "health and human rights" and the concept of "human rights impact assessment." Nancy Krieger et al., "Assessing Health Impact Assessment: Multidisciplinary and International Perspectives," *Journal Epidemiology Community Health* 57 (2003): 659–62.
26. Racial impact statements or assessments have been proposed in other contexts such as mitigating sentencing disparities. See, for example, Marc Mauer, "Racial Impact Statements As a Means of Reducing Unwarranted Sentencing Disparities," *Ohio State Journal of Criminal Law* 5 (2007): 19.
27. Nancy Krieger et al., "Assessing Health Impact Assessment: Multidisciplinary and International Perspectives," *Journal Epidemiology Community Health* 57 (2003): 659–62.

CONCLUSION

TOWARD A REMEDY FOR THE SOCIAL CONSEQUENCES OF RACIAL MYTHS

Kathleen Sloan

As the preceding chapters have amply documented and discussed, human societies have grouped and classified one another for millennia. Plato stratified societies into groupings of people with three distinct natural abilities to justify roles within a society. Other civilizations created concepts of royal bloodlines and hierarchies based on religious views, political power, and origin myths. Among these divisions and categories, none has been more powerful or enduring than the concept of race.

Beginning in eighteenth-century Europe, the idea of human "races" was accepted and given the imprimatur of scientific "fact" (see chapter 1). Historical developments of the time contributed significantly to this division of population groups centered on the idea of separate races because it greatly served the interests of the globally dominant European powers. Racial taxonomy was connected to ideologies of racial hierarchy, or the belief that some racial groups were superior to others. The racial hierarchies were rationalized by physical, social, cultural, and behavioral characteristics defined and measured by dominant groups. Racial taxonomies were perpetuated through legal and political policies, providing economic and cultural advantages for some races while hindering the advancement of others.

Competition between empires for land, human, and natural resources was played out on a global stage as the growth and development of shipbuilding and navies facilitated their expansion on a scale hitherto unknown. Colonial outposts were established in North America, Africa, East Asia, and the Indian subcontinent. This epic rivalry of conquest culminated in what Winston Churchill called the "first world war,"[1] referred to as the French and Indian War (1754–63) in North America (or "the war that made America")[2] and the Seven Years War in Europe (1756–63). From North America the war spread to Europe, then west Africa over control of the slave trade, and finally to India and the Philippines. It was an ultimate battle over which empire would prevail, Britain or France. When the war ended, Britain ruled more territory than had been possessed by the Roman Empire. The war was ultimately decisive in creating the British Empire and in determining the future of North America (along with the spread of the belief in separate races among the human species).

As Europeans fanned across the globe in a mission to acquire and control natural resources, they created belief systems to support their goals of colonization. These belief systems had the imprimatur of science and were based on visually apparent differences in physical appearance, reinforced by differences in cultural practices and worldviews. It was during this era of the eighteenth century that the idea of race was promulgated and provided with credibility by the scientists of the time. As Christina Snyder notes in her book *Slavery in Indian Country: The Changing Face of Captivity in Early America*, "Not until the late eighteenth century did Southern Indians begin to graft ideas about race onto their preexisting captivity practices."[3] Once the idea of not just separate races, but superior and inferior races, was established in the human psyche, practices such as genocide and slavery became acceptable, both logically and morally.

Awareness of this history is crucial to understanding how acceptance of the idea of race and the disparities that accompany it have led us to where we are today, when we face the formidable challenge involved in crafting solutions to this destructive legacy of racial injustice and the paradoxes of racial taxonomy, which, though refuted by science, remain strongly embedded in popular culture (see introduction). What began with colonial conquest and a global slave trade in the eighteenth century evolved into Manifest Destiny in the United States and shaped the development of the country and the attitudes of its inhabitants. In the nineteenth century, Andrew Jackson and the South Carolina statesman John C. Calhoun anticipated and shaped theories of scientific racism (the confluence of science, policy, and popular opinion) that gained credence in the 1830s; they argued that Indians were racial inferiors who had no place in the new American empire they sought to build.[4] As Snyder notes, the developing pseudoscience of phrenology, which supposedly used cranial morphology to measure intelligence, bolstered theories of scientific racism. In this American context, she describes how unquestioned acceptance of a belief in unequal races, buttressed by scientific racism, was used to make government policy:

> Philadelphia physician Samuel Morton's influential 1839 study *Crania Americana* used phrenology to formulate an elaborated racial hierarchy—whites at the top, Indians in the middle, and Africans at the bottom. Echoing Indians' earlier theory of polygenesis and the already hardened racial attitudes of white frontierspeople, elite Americans concluded that racial differences were immutable, propelling each group toward its separate destiny: whites would rule the continent, people of African descent would be subservient, Indians would disappear. Or so they believed. In any case, gone were the days when policymakers sought to integrate "civilized" Indians into the republic. By the Jackson era, American expansion showed little regard for nonwhites who stood in the way. The Indian Removal Act, the increasingly harsh slave codes of Southern states, and new restrictions on free people of color in the North were emblematic of the new doctrine of race in America.[5]

As this book's contributors have conveyed, rather than questioning the validity of a social construct as products of their time, place, and culture, many scientists of the eighteenth, nineteenth, and early twentieth centuries sought to identify a biological component to "race" to conform to the larger society's prejudices. Stephen Jay Gould wrote in *The Mismeasure of Man*: "In assessing the impact of science upon eighteenth and nineteenth century views of race, we must first recognize the cultural milieu of a society whose leaders and intellectuals did not doubt the propriety of racial ranking—with Indians below whites, and blacks below everybody else."[6]

The historical record is replete with documentation of the destruction that this impulse to divide has produced. The organization of patriarchal societies based on hierarchy, the acquisition of power, and the assertion of control requires classifications of "others" to divide and conquer. This need exploits a primitive ego-based desire in the human psyche for belonging for purposes of identity, security, and survival.

Societal organization based on hierarchy naturally is compelled to identify an "Other" for self-justifying manipulation, abuse, and control in order to create categories of superior and inferior (see chapter 2). Rather than "we are all in this together for the greatest good," an "us versus them" mentality evolves. Once established in the collective consciousness, any type of inequality and ill treatment becomes acceptable, up to and including murder, slavery, and genocide.

Historical examples can be found in the Christian crusading armies' use of the term "infidels" to refer to Muslim populations of the Middle East, embedding the concept of an alien, lesser human "Other." This phenomenon is readily used by powerful hierarchies to maintain and rationalize control. Among the most current is its employment by militaries during wartime. In this context, the "enemy" becomes the "Other" and therefore less than civilized, justifying subjugation and, in some cases, annihilation. This language was used by American forces in World War II against Japan, during the Korean War against Koreans, and against the Vietnamese during the war in Vietnam. Racial stereotyping in war continues with the US invasion and occupation of Iraq and the use of the single name "Hajji" to group together an entire population, collectively referring to them in a derogatory and dehumanizing manner.

It is relevant and important to provide the broader context to illustrate how pervasive the phenomenon of racial classification is and how necessary and difficult a transformation of thinking it is to reverse the myriad manifestations of this destructiveness (as Robert Pollack so effectively articulates in chapter 2 in his discussion of the hardening of synaptic connections in the brain over time).

At its most pragmatic and pervasive level, this remedial effort should begin not by denying but by seeking to understand how race has become a socially constructed reality. As a contributor to this volume (see chapter 12), Osagie Obasogie discusses the idea of race impact assessments (RIA), used as policy instruments in areas ranging from the criminal justice system to medicine, and illustrates how they constitute the most sensible and demonstrably effective solution. Similar to environmental impact assessments or health impact assessments, race impact assessments evaluate the possible impact, positive or negative, that a proposed policy or project may have on racial disparities in society. The International Association for Impact Assessment (IAIA) defines an impact assessment as "the process of identifying, predicting, evaluating and mitigating the social and other relevant effects of policy or project proposals prior to major decisions being taken and commitments made."[7] The purpose is to determine whether there is any adverse impact on different ethnic groups and if so to take appropriate action to prevent it. It is a legal process designed to embed equal opportunities in policy and is required where a policy is considered to have a bearing on race relations. In the United Kingdom, for example, under the Race Relations (Amendment) Act 2000, all public authorities have a duty to ensure their polices do not discriminate against any race or ethnic minority and that they promote equality of opportunity and good relation between people of different racial groups. Where a policy is considered to have a bearing on race relations, the public authority is required to undertake a Race Equality Impact Assessment (REIA).[8] An RIA can be a vital tool for preventing institutional racism and for identifying new options to remedy long-standing inequalities. The persistence of deep racial disparities and divisions across society is evidence of institutional racism—the routine, often invisible and unintentional production of inequitable social opportunities and outcomes. When racial equity is not consciously addressed, racial inequality is often unconsciously replicated. The use of RIAs in the United States is relatively new and still somewhat limited, but interest and initiatives are on the rise. The United Kingdom has been using them with success for nearly a decade.

The urgency of the need for creation of race impact assessments is demonstrated in today's newspapers and blogs and displayed across television and computer screens carrying coverage of Arizona's passage of a blatantly race-based anti-immigration law. In the guise of trying to stop the inflow of undocumented immigrants, the state passed a draconian law that essentially legalizes racial profiling.[9] SB 1070 is a radical anti-immigrant piece of legislation that will open the floodgates to racial profiling and abuses of civil liberties. The law will be challenged in court for both violating individual rights and being an illegal assertion of state authority given the federal government's primary responsibility for border and immigration matters. But in the meantime, the effects of its implementation will be sweeping, since the law legalizes racial profiling. State and local government law enforcement officers are required to determine if a person is illegally in the United States based on a "reasonable suspicion," an open-ended approach that will encourage suspicions based on race. The law does little, if anything, to prohibit police officers from relying on race or ethnicity in deciding whom to investigate.

Offering proof that as a society we have not come so far as we may wish to delude ourselves into believing, Bob Cesca writes in his blog on the *Huffington Post*:

> While there's no evidence that the Arizona law will feed the rise of a new underground forced labor market in the United States, it's clear that the various components of neo-slavery are here now. And the optics and civil liberties violations, to say nothing of the long-term consequences, are horrible. We're on the brink of rounding up Hispanic people on ridiculous charges while American corporations are actively engaged in the trafficking of illegal immigrant labor. How soon, I wonder, until we read about Hispanic people, citizens or otherwise, being picked up for not having their birth certificates and other "papers" in their back pockets and consequently shipped off on some sort of prison "work release" program to a cabbage farm or meat packing plant? Free labor is slave labor.[10]

As the world, and American society in particular, grapples with these painful truths, social justice advocates demand change and the academic world of science confronts a fundamental question: will it contribute to the reification of a concept openly discredited by its own discipline or will it insist upon fidelity to its defining characteristics as a branch of knowledge or study dealing with a body of facts or truths? It is the hope of the Council for Responsible Genetics, the contributing authors, and the coeditors of this book that the latter impulse prevail and that this volume will serve to inspire all those it reaches to see beyond the myths of social construction.

NOTES

1. H. V. Bowen, *War and British Society, 1688–1815* (Cambridge: Cambridge University Press, 1998), 7.
2. War That Made America Productions, LLC, and French and Indian War 250, Inc., "The War that Made America: The Story of the French and Indian War" (PBS Home Video, 2006).
3. Christina Snyder, *Slavery in Indian Country: The Changing Face of Captivity in Early America* (Cambridge, MA: Harvard University Press, 2010), 5.
4. Robert V. Remini, *Andrew Jackson and His Indian Wars* (New York: Viking, 2001), 7–20.
5. Snyder, *Slavery in Indian Country*, 235.
6. Stephen Jay Gould, *The Mismeasure of Man* (New York: Norton, 1981), 31.
7. International Association for Impact Assessment, *Principles of Impact Assessment Best Practice*, 1999, www.iaia.org
8. Department for Children, Schools and Families, *Bill Rammell Publishes Race Impact Assessment and Announces New Measures*, United Kingdom, March 26, 2007, www.dcsf.gov.uk.
9. State of Arizona, Senate, 49th Legislature, Second Regular Session. *Senate Bill 1070*. Phoenix, 2010, p. 1.
10. Bob Cesca. *Arizona Immigration Law Conjures Ghosts of Southern Neo-Slavery*, www.huffingtonpost.com, April 29, 2010.

CONTRIBUTORS

TROY DUSTER, PH.D., is Silver Professor of Sociology at New York University, and he also holds an appointment as Chancellor's Professor at the University of California, Berkeley. From 1996 to 1998, he served as member and then chair of the joint National Institutes of Health/Department of Energy advisory committee on Ethical, Legal, and Social Issues in the Human Genome Project (The ELSI Working Group). He is past president of the American Sociological Association (2004–05), and in 2003–04 served as chair of the Board of Directors of the Association of American Colleges and Universities. He is the former director of the American Cultures Center and the Institute for the Study of Social Change, both at the University of California, Berkeley.

DUANA FULLWILEY, PH.D., is an anthropologist of science and medicine whose research explores how personal identity, health status, and molecular genetic findings increasingly intersect. Currently, she is completing a book called *The Enculturated Gene*, which draws on ethnographic fieldwork in the United States, France, Senegal, and West Africa on locally varied versions of sickle cell science and disease embodiment. Since 2003, she has also conducted multisited field research in the United States on emergent technologies that measure human genetic diversity among populations and between individuals. In its detail, her work in the United States explores how political concepts of diversity, usually glossed as "race," function in genetic recruitment protocols and studies designs for research on complex diseases, "tailored medicine," and personal genomics. As an outgrowth of this research she also examines the cultural underpinnings of genetic ancestry testing and forensic science.

JOSEPH L. GRAVES, JR., PH.D., is Professor of Biological Sciences, North Carolina A & T State University. He is a Fellow of the Council of the American Association for the Advancement of Science (AAAS), Section G: Biological Sciences. His research concerns the evolutionary genetics of postponed aging and biological concepts of race in humans. He is the author of *The Emperor's New Clothes: Biological Theories of Race at the Millennium*

(Rutgers University Press, 2005) and *The Race Myth: Why We Pretend Race Exists in America* (Dutton Press, 2005). He presently serves as a member of the editorial board of Evolution: Education and Outreach, and he is chair of the Senior Advisory Board for the National Evolutionary Synthesis Center (NESCent.) Dr. Graves is included in C. W. Carey, *African Americans in Science: An Encyclopedia of People and Progress*, ABC-CLIO, 2008.

ELENA L. GRIGORENKO, PH.D., received a Ph.D. in general psychology from Moscow State University, Russia, in 1990, and a Ph.D. in developmental psychology and genetics from Yale University in 1996. Currently, she is Associate Professor of Child Studies, Epidemiology and Public Health, and Psychology at Yale and Adjunct Professor of Psychology at Moscow State University, Russia. Prof. Grigorenko has published more than 250 peer-reviewed articles, book chapters, and books. She is the recipient of multiple professional awards. Prof. Grigorenko's current research includes studies of cognitive and linguistic adaptation of international adoptees in the United States; learning disabilities in harsh developmental environments and their relation to infection, intoxication, and poverty in Africa; genes involved in language disorders in a genetically isolated population; genes involved in learning disabilities and cognitive processing, with special emphasis on studying minority samples in the United States; and interactions between genetic and environmental risk factors for conduct problems and the role of these factors in response to interventions in juvenile detainees.

JONATHAN KAHN, PH.D., J.D., is Professor of Law at Hamline University School of Law. He holds a Ph.D. in History from Cornell University and a J.D. from Boalt Hall School of Law. He writes on issues in history, politics, and law and specializes in biotechnology's implications for our ideas of identity, rights, and citizenship. His book *Race in a Bottle: Law, Commerce and the Uses of Race in Biomedicine* is forthcoming from Harvard University Press.

KENNETH K. KIDD, PH.D., is a Professor of Genetics at Yale University. He was elected a Fellow of the American Association for the Advancement of Science, was a Mellon Distinguished Visiting Professor at the University of Witwatersrand, South Africa, and was awarded the Carter Medal and Lecture from the British Society of Medical Genetics. He has won the "Profiles in DNA Courage" award from the US National Institute of Justice, and the "Biomedical Paper of the Year" award in 2002, given by *The Lancet*. Prof. Kidd has published widely in many areas of human genetics. His research interests include Complex Human Disorders including Neuropsychiatric Disorders; his current research is focused on Human Population Genetics and Evolution.

OSAGIE K. OBASOGIE, PH.D., J.D., is an Associate Professor of Law at the University of California, Hastings, with a joint appointment at the University of California, San Francisco, Department of Social and Behavioral Sciences, and is also a Senior Fellow at the Center for Genetics and Society. Obasogie's scholarly interests include Constitutional law, bioethics, sociology of law, and reproductive and genetic technologies. His writings have spanned both academic and public audiences, with journal articles in the *Law & Society Review*, *Yale Journal of Law and Feminism*, *Stanford Journal of Civil Rights and Civil Liberties*, and the *Journal of Law, Medicine, and Ethics*, along with commentaries in outlets including the *Los Angeles Times*, *Boston Globe*, *San Francisco Chronicle*, and *New Scientist*. Obasogie received his B.A. with distinction from Yale University, his J.D. from Columbia Law School where he was a Harlan Fiske Stone Scholar, and his Ph.D. in Sociology from the University of California, Berkeley, where he was a fellow with the National Science Foundation.

PILAR N. OSSORIO, PH.D., J.D., is Associate Professor of Law and Bioethics at the University of Wisconsin at Madison, and Program Faculty in the Graduate Program in Population Health at the UW. Prior to taking her position at UW, she was Director of the Genetics Section at the Institute for Ethics at the American Medical Association, and taught as an adjunct faculty member at the University of Chicago Law School. Dr. Ossorio received her Ph.D. in Microbiology and Immunology in 1990 from Stanford University. She went on to complete a postdoctoral fellowship in cell biology at Yale University School of Medicine. In 1993 she served on the Ethics Working Group for President Clinton's Health Care Reform Task Force. In 1994 she took a position with the Department of Energy's program on the Ethical, Legal, and Social Implications of the Human Genome Project. Dr. Ossorio received her J.D. from the University of California at Berkeley School of Law in 1997, where she was elected to the legal honor society Order of the Coif. She is a member of the National Advisory Council for Human Genome Research, and she advises on several large-scale international genomics projects.

ROBERT E. POLLACK, PH.D., is professor of biological sciences, member of the faculty of the Earth Institute, lecturer in psychiatry at the Center for Psychoanalytic Training and Research, adjunct professor of science and religion at Union Theological Seminary, Director of University Seminars (2011), and Director of the Center for the Study of Science and Religion at Columbia University. Dr. Pollack graduated from Columbia University with a B.A. in physics, and received a Ph.D. in biology from Brandeis University. He has been a professor of biological sciences at Columbia since 1978, and was dean of Columbia College from 1982 to 1989. He received the Alexander Hamilton Medal from Columbia University and has held a Guggenheim Fellowship. He currently is on the advisory boards of Columbia/Barnard Hillel and has been a Senior Consultant for the Director, Program of Dialogue on Science, Ethics, and Religion, American Association for the Advancement of Science (AAAS). He is a Fellow of the AAAS, and the World Economic Forum in Davos. He is the author of *Signs of Life: The Languages and Meanings of DNA* (Houghton Mifflin/Viking Penguin, 1994), *The Missing Moment: How the Unconscious Shapes Modern Science* (Houghton Mifflin, 1999), and *The Faith of Biology and the Biology of Faith: Meaning, Order and Free Will in Modern Medical Science* (Columbia University Press, 2000). *Signs of Life* received the Lionel Trilling Award and has been translated into six languages.

MICHAEL T. RISHER, J.D., is a staff attorney at the American Civil Liberties Union (ACLU) of Northern California, the nation's largest ACLU affiliate, where he focuses on freedom of speech, open government, criminal justice, and privacy rights. He is currently litigating a federal challenge to California's law mandating DNA collection from anyone arrested on suspicion of a felony for inclusion in the FBI's national forensic DNA database, the Combined DNA Index System (CODIS). Before joining the ACLU-NC, Risher was a Deputy Public Defender in Alameda County from 1998 to 2005. He has also served as the legal affairs advisor for the Lindesmith Center, a drug-policy think tank, and as a clerk to Judge Karen Nelson Moore on the US Court of Appeals. He graduated with honors from Harvard College and with distinction from Stanford Law School, where he was elected to the Order of Coif.

STEVEN E. STEMLER, PH.D., is an Assistant Professor of Psychology at Wesleyan University. He received his doctorate in Educational Research, Measurement, and Evaluation from Boston College, where he worked at the Center for the Study of Testing, Evaluation, and Educational Policy and the TIMSS International Study Center. Prior to joining the faculty at Wesleyan, Professor Stemler spent four years at Yale University, where he was an Associate Research Scientist in the Department of Psychology. His areas of interest and expertise include measurement and assessment, particularly within the domains of intelligence, intercultural literacy, and ethical reasoning.

ROBERT J. STERNBERG, PH.D., is Provost and Senior Vice President at Oklahoma State University. Prior to that, he was Dean of the School of Arts and Sciences and Professor of Psychology and Education at Tufts University. Sternberg is a former president of the American Psychological Association and Eastern Psychological Association and is President of the International Association of Cognitive Education and Psychology. His Ph.D. is from Stanford and he holds eleven honorary doctorates. A member of the American Academy of Arts and Sciences, Dr. Sternberg's research covers a wide range of areas, including intelligence, creativity, wisdom, leadership, love and close relationships, and hate. He is the author of over 1200 books and articles, and has won roughly two dozen awards for his scholarship. His research has taken him to five different continents, where he has studied the relationship between culture and competence.

HELEN WALLACE, PH.D., is Director of GeneWatch UK, a nonprofit science policy research group that aims to ensure that genetics is used in the public interest. The focus of her work has been on the assessment and regulation of genetic tests and genetic databases, including the genetic research project UK Biobank and the police National DNA Database. Helen has published and spoken widely about the issues associated with the rapid expansion of DNA collection and retention in Britain and has considerable experience in political and media work. Dr. Wallace has a degree in physics and a Ph.D. in applied mathematics (environmental modeling) and has worked as an environmental scientist for a consultancy firm and for Greenpeace UK.

PATRICIA J. WILLIAMS, J.D., a Professor of Law at Columbia University, holds a B.A. from Wellesley College and a J.D. from Harvard Law School. She was a fellow in the School of Criticism and Theory at Dartmouth College and has been an Associate Professor at the University of Wisconsin School of Law and its Department of Women's Studies. Williams also worked as a consumer advocate in the Office of the City Attorney in Los Angeles. A member of the State Bar of California and the Federal Court of Appeals for the Ninth Circuit, Williams has served on the advisory council for the Medgar Evers Center for Law and Social Justice of the City University of New York and on the board of governors for the Society of American Law Teachers, among others. Her publications include *Anthony Burns: The Defeat and Triumph of a Fugitive Slave*, *On Being the Object of Property*, *The Electronic Transformation of Law*, and *We Are Not Married: A Journal of Musings on Legal Language and the Ideology of Style.* In 1993, Harvard University Press published Williams's *The Alchemy of Race & Rights* to widespread critical acclaim. She is also author of *The Rooster's Egg* (Harvard, 1995), *Seeing a Color-Blind Future: The Paradox of Race* (Noonday Press, 1998), and, most recently, *Open House: On Family Food, Friends, Piano Lessons and The Search for a Room of My Own* (Farrar, Straus and Giroux, 2004).

MICHAEL YUDELL, PH.D., M.P.H., is an Assistant Professor in the Department of Community Health and Prevention at the Drexel University School of Public Health. Yudell received his Ph.D. in Sociomedical Sciences from Columbia University, an M.P.H. from the Mailman School of Public Health at Columbia University, an M.Phil. from the City University of New York in US History, and a B.A. from Tufts University. Yudell is the author, with Rob DeSalle, of *Welcome to the Genome: A User's Guide to the Genetic Past, Present, and Future* (John Wiley and Sons, 2004). Yudell and DeSalle also edited *The Genomic Revolution: Unveiling The Unity Of Life* (National Academies of Science Press, 2002). He is currently completing the book *Making Race: Biology and the Evolution of the Race Concept in Twentieth Century American Thought*, a history that examines the way in which biologists and geneticists shaped the race concept during the twentieth century, from eugenics to the sequencing of the human genome. His next project is a history of autism spectrum disorders.

EDITORS

SHELDON KRIMSKY, PH.D., received his Ph.D. in philosophy from Boston University and a Master's degree in physics from Purdue University. He is a professor in the Department of Urban & Environmental Policy & Planning, School of Arts & Sciences, and Adjunct Professor of Public Health and Community Medicine, School of Medicine, Tufts University. His research has focused on the linkages between science/technology, ethics/values, and public policy. He is the author of nine books, which include *Genetic Alchemy* (1982), *Biotechnics and Society: The Rise of Industrial Genetics* (1991), *Hormonal Chaos: The Scientific and*

Social Origins of the Environmental Endocrine Hypothesis (2000), and *Science in the Private Interest: Has the Lure of Profits Corrupted Biomedical Research?* (2003). His coauthored book *Genetic Justice: DNA Databanks, Criminal Investigations and Civil Liberties* was published by Columbia University Press (2011). Professor Krimsky served on the National Institutes of Health's Recombinant DNA Advisory Committee, was a consultant to the Presidential Commission for the Study of Ethical Problems in Medicine and Biomedical and Behavioral Research and to the Congressional Office of Technology Assessment on its series of reports on biotechnology, was chairperson of the Committee on Scientific Freedom and Responsibility for the American Association for the Advancement of Science (AAAS), and was elected Fellow of the AAS for "seminal scholarship exploring the normative dimensions and moral implications of science in its social context." Professor Krimsky is a founding member of the Council for Responsible Genetics and serves on its Board.

KATHLEEN SLOAN is a communications, public relations, and marketing professional and social justice advocate with over twenty years of experience. She holds a Master's degree in International Relations and has lived and worked in Russia, Hungary, and Ireland. She is the National Organization for Women's (NOW) Main Representative at the United Nations (UN) and is a member of its Board of Directors. She has directed nonprofit organizations and was the Public Relations Director of a multinational corporation with operations in eighty countries and over eight hundred employees. She conducts advocacy for and writes on women's, indigenous peoples', and human rights issues. She was awarded a Cochran Fellowship by the US Department of Agriculture (USDA), promoting Russian-American cooperation. She currently is a US State Department Consultant on women's issues for its International Information Program at embassies and consulates around the world.

INDEX